ENZYME KINETICS

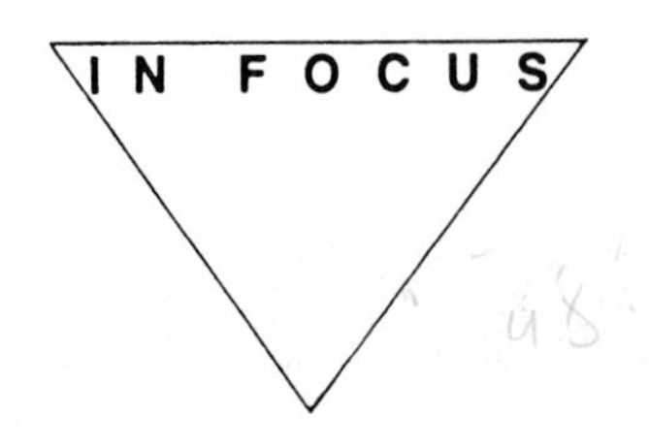

Series editors

David Rickwood

Department of Biology, University of Essex, Wivenhoe Park, Colchester, Essex CO4 3SQ, UK

David Male

Institute of Psychiatry, De Crespigny Park, Denmark Hill, London SE5 8AF, UK

Titles published in the series:

Antigen-Presenting Cells

B Lymphocytes

***Complement**

Enzyme Kinetics

Gene Structure and Transcription

***Immune Recognition**

***Lymphokines**

Membrane Structure and Function

Molecular Basis of Inherited Disease

Regulation of Enzyme Activity

***Published in association with the British Society for Immunology.**

ENZYME KINETICS

Athel Cornish-Bowden

CNRS–CBM, 31 Chemin Joseph Aiguier, 13402 Marseilles Cedex 09, France

Christopher W. Wharton

Department of Biochemistry, University of Birmingham, UK

Published by:
IRL Press Limited
PO Box 1,
Eynsham,
Oxford OX8 1JJ,
UK

First published 1988
Reprinted 1990

British Library Cataloguing in Publication Data

Cornish-Bowden, Athel, *1943-*
Enzyme kinetics.—(In focus).
1. Enzymes. Chemical reactions. Kinetics
I. Title II. Wharton, Christopher W.
III. Series
547.7′5804594

ISBN 1 85221 074 5

Printed by Information Press Ltd, Oxford, England.

Preface

To describe the whole of enzyme kinetics in the compass of a relatively short book would clearly be an impossible task, and we have not attempted it. The serious student of enzymes will need to look to more advanced texts. Nonetheless, there is a core of enzyme kinetics that every biochemist needs to be familiar with, and every student of biochemistry needs to be taught; it is this core that we have tried to describe in this book. We believe that there is nothing here that is inappropriate in an undergraduate course, though not all of the topics are elementary. This is especially true of the last chapter, where we have tried to show that despite changes in fashion there is still plenty of life in enzymology, and exciting work still to be done.

We are grateful to Marilú Cárdenas for helpful comments.

Athel Cornish-Bowden
Christopher W.Wharton

Dedications

To Isadora

To Amanda and Debbie

Contents

4. Inhibition of enzyme activity

5. The pH-dependence of enzyme-catalysed reactions

6. Enzyme mechanisms

Abbreviations and symbols

a_0	total enzyme concentration
A (or B)	substrate
Ala	alanine
Asp	aspartate
Cys	cysteine
dx/dt	rate of change of EA concentration
e_0	total enzyme concentration
E	enzyme
EA(B)	enzyme – substrate complex
EAI	enzyme – substrate – inhibitor complex
Glu	glutamate
Gly	glycine
h	hydrogen ion concentration
His	histidine
I	inhibitor
Ile	isoleucine
IR	infrared
k_0	catalytic constant
k_A	specificity constant
k_n	rate constant
k_{obs}	observed rate constant
K_a	acid dissociation constant
K_{eqm}	equilibrium constant
K_i	competitive inhibitor constant
K_m	Michaelis constant
K_m^{app}	apparent Michaelis constant
K_s	equilibrium dissociation constant
Lys	lysine
NMR	nuclear magnetic resonance
p	concentration of product
P (or Q)	product
Pro	proline
Ser	serine
Thr	threonine

Tyr	tyrosine
UV	ultraviolet
v	rate
V	limiting rate
Val	valine
x	concentration of EA

1

Simple enzyme kinetics

1. Enzyme saturation

The kinetic feature that most distinguishes enzyme-catalysed reactions from simple chemical reactions is that they show *saturation*. Nearly all enzyme-catalysed reactions show a first-order dependence of rate on substrate concentration at very low concentrations, but instead of increasing indefinitely as the concentration increases, the rate approaches a limit at which there is no dependence of rate on concentration and the reaction becomes of *zero order* with respect to substrate. This behaviour, which was known from the earliest studies of enzymes, is illustrated in *Figure 1.1*. Note especially that, despite what is shown in some textbooks, the curve does not reach the limit at any finite concentration, but rather remains far from it at even the highest concentrations that can realistically be achieved. We shall consider the reasons and consequences of this below.

The first investigators to provide a reasonably clear interpretation of enzyme saturation were A.J.Brown and V.Henri, but the somewhat later work of Michaelis and Menten (1) is usually taken as the starting point when discussing enzyme kinetics. Their claim to be regarded as the founders of enzyme kinetics rests not so much on their interpretation of saturation as on the fact that they were the first to carry out experiments in a modern way, with proper control of pH, use of initial rates rather than whole time courses, and allowances for non-enzymic processes. Their mechanism supposes that the first step in the reaction is the binding of the substrate (A) to the enzyme (E) to form an *enzyme–substrate complex* (EA) which then reacts to give the product (P) with the regeneration of the free enzyme:

$$\begin{array}{ccccccccc} \mathrm{E} & + & \mathrm{A} & \underset{K_s}{=} & \mathrm{EA} & \xrightarrow{k_0} & \mathrm{E} & + & \mathrm{P} \\ e_0 - x & & a & & x & & & & p \end{array} \quad (1.1)$$

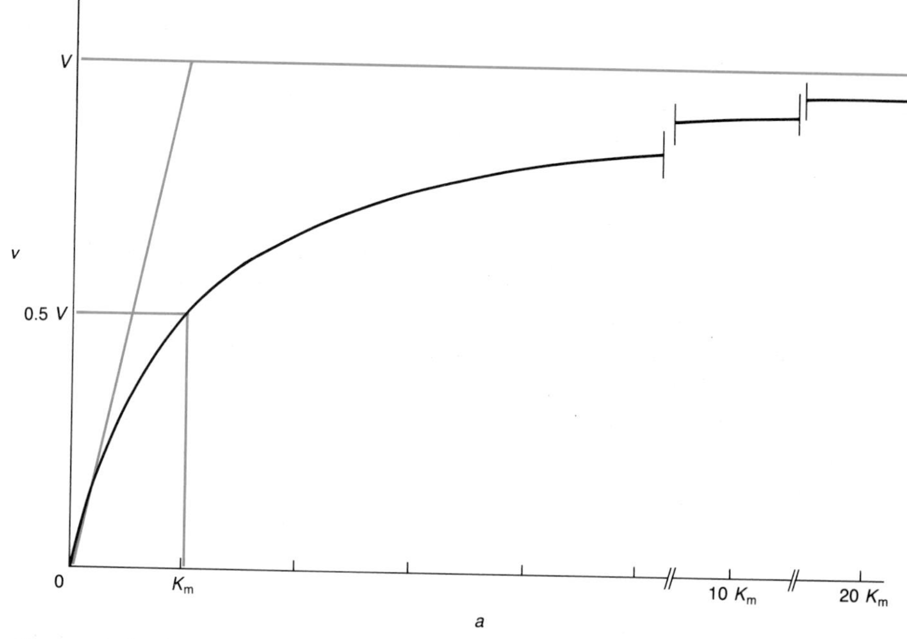

Figure 1.1. Dependence of rate v on substrate concentration a for a typical enzyme-catalysed reaction. The coloured additions illustrate the meanings of the limiting rate V and the Michaelis constant K_m.

If the *total* enzyme concentration is e_0 and the concentration of EA is x, the concentration of *free* enzyme must be e_0 minus x, as indicated under the equation, because all of the enzyme can only exist in one of the two forms. We can apply the same argument to the substrate, so that if the total substrate concentration is a_0 the free concentration must be a_0 minus x, at least until there has been sufficient time for the conversion of significant amounts of A into P. However, although it is possible to derive a rate equation in this way it is unnecessarily complicated because of the presence of square roots. Experimentally, one nearly always makes measurements with substrate concentrations that are very large compared with enzyme concentrations (e.g. 1 mM as compared with 1 nM), so that the analysis can be simplified by assuming that a_0 is so much larger than e_0 that a is effectively equal to a_0 minus x and so a is not significantly different from a_0. Hence, it is not necessary to distinguish between the free and total concentrations of substrate, and one can write both just as a. *In this book we shall always make this assumption*, even if it is not stated explicitly.

Michaelis and Menten assumed that the first step was an *equilibrium*, where equilibrium dissociation constant K_s equals [E][A]/[EA] which in turn is equal to $(e_0 - x)a/x$. This definition can easily be rearranged to express x in terms of e_0, a and K_s as follows:

$$x = e_0 a/(K_s + a) \tag{1.2}$$

Now, as product is released only in the second step of the reaction, the rate of the overall reaction is the rate of this step. As it is a first-order conversion of EA into E and P, with rate constant k_0 as indicated in the mechanism, the rate v is equal to k_0x or, taking into account Equation 1.2:

$$v = k_0e_0a/(K_s + a) \tag{1.3}$$

This, then, is a modern formulation of the equation obtained by Michaelis and Menten. Confusingly, however, it is *not* the equation known to modern biochemists as the 'Michaelis–Menten equation': this name is reserved for the steady-state equation (Equation 1.12) that we shall consider in Section 2.

Van Slyke and Cullen (2) considered the problem at about the same time as Michaelis and Menten, but in a different way. They also imagined a two-step process, but instead of treating the first as an equilibrium they treated it as irreversible, and they argued that the total time required could be treated as the sum of the times for the two steps:

$$\begin{array}{ccccc} \mathrm{E} + \mathrm{A} & \xrightarrow{k_1} & \mathrm{EA} & \xrightarrow{k_2} & \mathrm{E} + \mathrm{P} \\ e_0 - x \quad a & & x & & p \end{array} \tag{1.4}$$

At the instant of mixing enzyme with substrate, before any EA or P has been produced, all of the enzyme exists as free enzyme, and so the initial rate of the first step must be k_1e_0a. If this rate were maintained until all of the enzyme was converted into EA the time required would be e_0/k_1e_0a, or $1/k_1a$. If this conversion were then followed by the complete conversion of EA back into E at a rate k_2e_0, the time required would be e_0/k_2e_0, or $1/k_2$. In practice, of course, the two processes occur simultaneously, but for any one enzyme molecule they are consecutive, and we can still regard these times as the average times required to convert a single molecule of E into EA and to convert a molecule of EA into E. Thus the average time required to carry out both steps, with concomitant conversion of a molecule of A into P, is the sum of the two. It follows that the rate of reaction is the concentration of molecules able to perform the transformation divided by this total time:

$$v = e_0/[(1/k_1a) + (1/k_2)] \tag{1.5}$$

Arranged more conventionally, this is of the same form as Equation 1.3:

$$v = k_2e_0a/[(k_2/k_1) + a] \tag{1.6}$$

with k_2 replacing k_0 and k_2/k_1 replacing K_s. Replacing k_0 by k_2 is just a change in symbol for what is really the same quantity, the rate constant for conversion of EA into products. However, replacing K_s, an equilibrium constant, by k_2/k_1, the ratio of two rate constants for consecutive reactions, is more than just a change in symbol, as the two quantities have conceptually different meanings. However, this does not mean that they can be easily distinguished experimentally, as behaviour obeying Equation 1.3 can equally well be interpreted in terms of Equation 1.6. As we shall see in the next section, both equations are special cases of a more general treatment.

2. The steady-state assumption

Briggs and Haldane (3) showed that the equilibrium-binding mechanism of Michaelis and Menten and the irreversible-binding mechanism of Van Slyke and Cullen were special cases of a more general mechanism in which the binding of substrate was assumed to be reversible, but not necessarily at equilibrium during the reaction:

$$\underset{e_0 - x}{\mathrm{E}} + \underset{a}{\mathrm{A}} \underset{k_{-1}}{\overset{k_1}{\rightleftharpoons}} \underset{x}{\mathrm{EA}} \overset{k_2}{\longrightarrow} \mathrm{E} + \underset{p}{\mathrm{P}} \tag{1.7}$$

The rate of change of the concentration of intermediate ($\mathrm{d}x/\mathrm{d}t$), is the difference between the rate at which it is being produced from E and A and the sum of the rates at which it is being converted back into E and A and forward into E and P:

$$\mathrm{d}x/\mathrm{d}t = k_1(e_0 - x)a - k_{-1}x - k_2x \tag{1.8}$$

Briggs and Haldane postulated that although at the instant of mixing $\mathrm{d}x/\mathrm{d}t$ must be positive, because no EA would then exist at that instant, the rates of removal of EA would rapidly increase and, after a very short time, they would balance its rate of production. A *steady state* would be established, in which $\mathrm{d}x/\mathrm{d}t$ is zero, so

$$k_1(e_0 - x)a - k_{-1}x - k_2 = 0 \tag{1.9}$$

This may be rearranged to express x in terms of e_0 and a:

$$x = \frac{k_1e_0a}{k_{-1} + k_2 + k_1a} \tag{1.10}$$

The reaction rate is the rate of the step in which P is produced, that is k_2x, or

$$v = \frac{k_1k_2e_0a}{k_{-1} + k_2 + k_1a} \tag{1.11}$$

Although this is not immediately of the form of Equations 1.3 and 1.6 it may be written so that it is by dividing all terms by k_1, that is:

$$v = \frac{k_0 e_0 a}{K_m + a} \tag{1.12}$$

where k_2 is written as k_0 (for reasons that will be considered shortly) and $(k_{-1} + k_2)/k_1$ as K_m, the Michaelis constant. This is the *Michaelis–Menten equation*, the fundamental equation of enzyme kinetics. This name is *not* normally applied to the equation derived with the equilibrium assumption of Michaelis and Menten, but to the equation obtained with the steady-state assumption.

3. The Michaelis – Menten equation

The two-step mechanism shown in Equation 1.7 is only the simplest of an infinite range of mechanisms that give rate equations of the form of the Michaelis–Menten equation, Equation 1.12. Thus, although the two-step mechanism is useful for discussion, one cannot be certain that it is the true mechanism simply because one observes adherence to the Michaelis–Menten equation experimentally. It is therefore best to represent its parameters as in Equation 1.12, with symbols k_0 and K_m that do not imply that they refer to particular steps of a mechanism, even though in the two-step case k_0 is in fact identical to k_2.

Although k_0 may not refer to a single step of a mechanism, it does have the properties of a first-order rate constant, defining the capacity of the enzyme–substrate complex, once formed, to form the product, P. It is commonly called the *catalytic constant* of the enzyme (and often symbolized as k_{cat}); the alternative name *turnover number* is sometimes used, though it is becoming less common. Values of about 10^3 s^{-1} are typical for k_0, and some enzymes show much larger values. It is a useful measure of catalytic activity, but one cannot always measure it, especially in the early stages of characterizing an enzyme, because the enzyme concentration is often unknown or difficult to measure. For this reason $k_0 e_0$ is often replaced by V (or V_{max}), a quantity known as the *limiting rate*.

The other parameter of the Michaelis–Menten equation is called the *Michaelis constant* (K_m). It corresponds to K_s in Equation 1.3, and resembles it in having the dimensions of concentration, that is of a dissociation constant, and in specifying the relative concentrations of free enzyme, free substrate and enzyme–substrate complex. However, unlike an equilibrium constant, it defines these concentrations under steady-state reacting conditions, not at equilibrium. Its value, K_m, is equal to $(k_{-1} + k_2)/k_1$ and it approximates to k_{-1}/k_1, that is to the equilibrium dissociation constant K_s, only if k_{-1} is large compared with k_2; it approximates to k_2/k_{-1}, the corresponding constant in Equation 1.6, if k_2 is large compared with k_{-1}.

Although it has often been suggested that K_m can in practice be assumed to be similar in magnitude to K_s, there is no good reason for this to be true and it is much more likely that enzymes will evolve with values of k_2 that are larger than k_{-1}. This is because once the enzyme–substrate complex is formed it is in the interests of catalytic efficiency to convert it into products as fast as possible (cf. Chapter 6, Section 5).

If the substrate concentration a is much smaller than K_m, it can be ignored by comparison with K_m in Equation 1.12, which thus simplifies to:

$$v \simeq (k_0/K_m)e_0a \tag{1.13}$$

that is with first-order dependences on both the enzyme and substrate, or second-order kinetics overall. It is evident that k_0/K_m is an important quantity and is more than just a ratio of two other quantities. It is sometimes called the 'second-order rate constant' for the reaction, but as it is the parameter that defines enzyme specificity, as we shall discuss in Chapter 4, Section 5, it is also known as the *specificity constant* and symbolized as k_A, where the subscript specifies which substrate is considered.

As a increases it becomes similar in magnitude to K_m, surpasses it, and eventually makes it insignificant. In the limit at high substrate concentrations, therefore, Equation 1.12 simplifies to:

$$v \simeq k_0e_0 = V \tag{1.14}$$

This is the reason for giving V the name 'limiting rate'. It is not reached at any finite value of a and is, moreover, approached rather slowly, so that, for example when a is 10 times the K_m then v is only 0.91 V, still 9% less than V. There is no maximum value of v in the mathematical sense, that is, there is no point at which the slope of the plot of v against a is zero, and thus the name 'maximum rate' (or 'maximum velocity') is not an accurate description, though it is very common.

The concentration at which an enzyme obeying Michaelis–Menten kinetics is 'half-saturated', that is, where v is equal to 0.5 V, may readily be shown to be when a is equal to K_m by substituting this value of a in Equation 1.12. This emphasizes that the Michaelis constant is a concentration, and gives physical meaning to it as the half-saturation concentration. It is often believed that the K_m value for an enzyme with its natural substrate is similar to the physiological concentration of the substrate. Substrate concentrations *in vivo* are difficult to measure and so there is no extensive evidence for the truth or otherwise of this supposition.

It follows that the plot of v against a has the shape shown in *Figure 1.1*, that is, the experimental behaviour observed for many enzymes is, in fact, explained by the Briggs–Haldane mechanism.

Although the Michaelis–Menten equation is usually written as Equation 1.12, that is, in terms of k_0 (or V) and K_m, there is nothing fundamental or necessary about this choice (which derives more from history than anything else), and one can equally legitimately write it in terms of k_0 and k_A:

$$v = k_0 k_A e_0 a/(k_0 + k_A a) \tag{1.15}$$

or in terms of k_A and K_m:

$$v = k_A e_0 a/(1 + a/K_m) \tag{1.16}$$

In many ways it would be more convenient if the study of enzymes had developed differently and Equation 1.15 were the most familiar form of the Michaelis–Menten equation. This would simplify discussion of numerous aspects of the subject, such as effects of inhibitors, estimation of parameters by graphical or statistical means, etc. In this book, however, we shall follow the most familiar practice of writing it as Equation 1.12, though we shall stress the importance of the specificity constant several times.

4. The validity of the steady-state assumption

As the steady-state assumption is fundamental in much of enzyme kinetics we should pause to enquire whether it is readily valid, which we can do by integrating Equation 1.8 without making the steady-state assumption. Strictly this is impossible, because a is a variable (as A is being converted into P), but if the steady state is established so rapidly that the amount of product formation during the early stages is trivial we can treat a as a constant. Then the variables in Equation 1.8 can readily be separated:

$$\int \frac{\mathrm{d}x}{k_1 e_0 a + (k_1 a_0 + k_{-1} + k_2)x} = \int \mathrm{d}t \tag{1.17}$$

and integrated, to yield:

$$x = \frac{k_1 e_0 a\{1 - \exp[-(k_{-1} + k_2 + k_1 a)t]\}}{k_{-1} + k_2 + k_1 a} \tag{1.18}$$

in which the constant of integration has been evaluated by putting x to zero when t is zero (because no enzyme–substrate complex can exist before any time has elapsed for it to be formed).

Despite its more complex appearance this is simply the steady-state expression for x contained in Equation 1.10 multiplied by the factor in curly brackets, which has a value of zero when t is zero, but rapidly approaches unity as t increases. As the coefficient of $-t$ in the exponential term typically has a value of at least 1000 s^{-1}, a simple calculation shows that the value of x given by Equation 1.18 exceeds 99.9% of that given by Equation 1.10 within a few milliseconds.

This result justifies our initial assumption that a could be treated as a constant, because steady-state measurements are normally done with enzyme concentrations small enough that the time scale of the experiment is of the order of minutes or more, with insignificant conversion of substrate into product during the first 10 ms.

Although this discussion has been in the context of the simplest model for Michaelis – Menten kinetics, and although the time required to establish a steady state will normally be longer for more complex mechanisms, it is still normally short compared with the time scale of the experiment. If one wishes to study the events leading up to the establishment of the steady state one must use special equipment and techniques suitable for making observations on a time scale of milliseconds or shorter (see Section 8).

5. Plots of the Michaelis – Menten equation

The most natural way of plotting steady-state kinetic data is to plot the rate v against the substrate concentration a, as in *Figure 1.1*, and this is certainly the clearest way to display the behaviour. However, it is not a good way to determine the kinetic parameters, because the line is curved and it does not approach the limit fast enough for one to be able to judge accurately where the limit is. For many years, therefore, biochemists have preferred to transform the Michaelis – Menten equation into the equation for a straight line. There are three ways of doing this, the least satisfactory of which has, unfortunately, proved to be by far the most popular. This is obtained by taking reciprocals of both sides of Equation 1.12:

$$\frac{1}{v} = \frac{1}{V} + \frac{K_m}{V} \cdot \frac{1}{a} \tag{1.19}$$

or

$$\frac{e_0}{v} = \frac{1}{k_0} + \frac{1}{k_A} \cdot \frac{1}{a} \tag{1.20}$$

The second of these equivalent representations emphasizes the correspondence between k_0 and k_A, which is somewhat obscured by the usual choice of K_m and V as the parameters of the Michaelis – Menten equation. Either way, a plot of $1/v$ against $1/a$ is a straight line, with a slope of $1/k_A e_0$ which is equal to K_m/V, an intercept on the ordinate ($1/v$) axis of $1/k_0 e_0$ which is equal to $1/V$, and an intercept on the abscissa ($1/a$) axis of $-k_A/k_0$ equal to $-1/K_m$ (*Figure 1.2*).

This plot is very widespread in the literature and any biochemist should be able to recognize and interpret it. Nonetheless, *we cannot emphasize too strongly that we do not advise its use for determining kinetic parameters. The time to learn good practices is now*! The objection to the *double-reciprocal plot*, as it

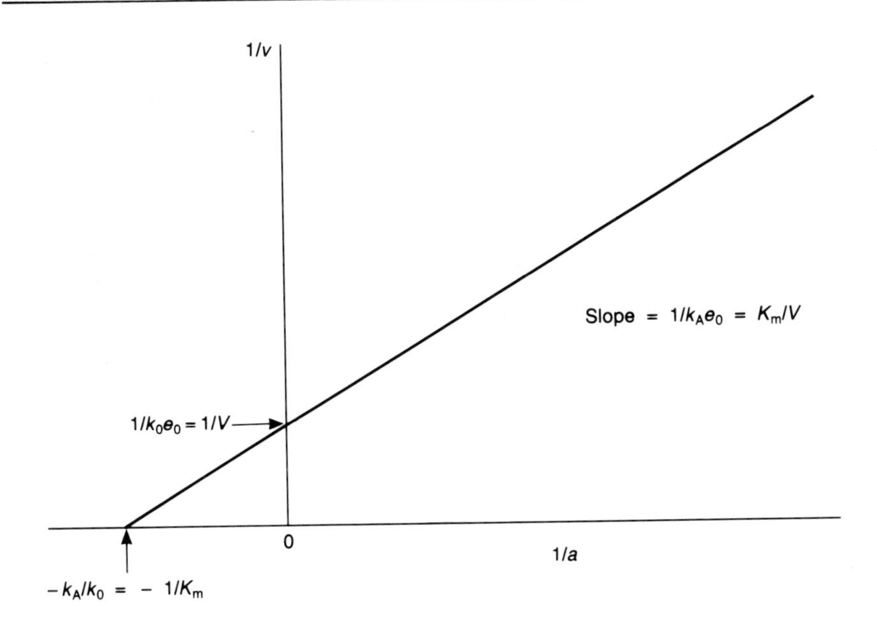

Figure 1.2. The double-reciprocal plot of reciprocal rate $1/v$ against reciprocal substrate concentration $1/a$.

is commonly known, is that it distorts the appearance of any experimental error in the primary observations of v, so that one cannot judge which points are most accurate when drawing a straight line through the set of points. This may be illustrated by a simple calculation. If a true rate v is 1 μM s^{-1}, but is in error by 0.1 μM s^{-1}, so that the measured v is 1.1 μM s^{-1}, then the observed $1/v$ is in error by -0.091 s μM^{-1}: 0.909 s μM^{-1} instead of 1.0 s μM^{-1}. Consider now a second measurement with a true rate five-fold higher, 5 μM s^{-1}, but subject to the same error of 0.1 μM s^{-1}, so that the observed rate is 5.1 μM s^{-1}; now the observed $1/v$ is in error by 0.0039 s μM^{-1}: 0.1961 s μM^{-1} instead of 0.2 s μM^{-1}. Thus equal errors in v are reflected in errors in $1/v$ differing by a factor of 25! As one cannot properly judge the amount of weight to be given to the different points by eye, one cannot obtain accurate results from a double-reciprocal plot. This argument is independent of any assumptions about whether all v values are truly subject to similar errors, whether measured in percent or in μM s^{-1}; it is simply that when the plot distorts the error behaviour so grotesquely one cannot make an accurate judgement about it.

Neither of the other straight-line plots is entirely free from distortion, but both are less severely affected than the double-reciprocal plot. The first of these can be generated by multiplying both sides of Equations 1.19 and 1.20 by a:

$$\frac{a}{v} = \frac{K_m}{V} + \frac{1}{V} \cdot a \tag{1.21}$$

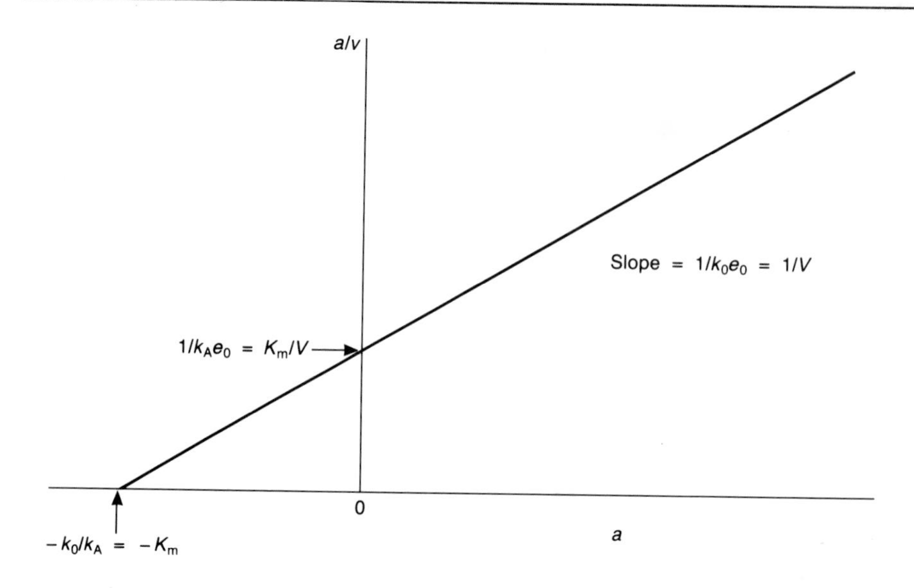

Figure 1.3. The plot of a/v against a.

or

$$\frac{e_0 a}{v} = \frac{1}{k_A} + \frac{1}{k_0} \cdot a \qquad (1.22)$$

Thus a plot of a/v against a is also a straight line (*Figure 1.3*). Now the slope, $1/k_0e_0$, is equal to $1/V$; the ordinate intercept, $1/k_Ae_0$, is equal to K_m/V; and the abscissa intercept, $-k_0/k_A$, is equal to $-K_m$. Note that the slope and ordinate intercept are interchanged from those of the double-reciprocal plot.

The equation for the third straight-line plot is obtained by multiplying both sides of Equation 1.21 by Vv/a and rearranging:

$$v = V - K_m\,(v/a) \qquad (1.23)$$

Thus a plot of v against v/a is also a straight line (*Figure 1.4*). The slope is $-K_m$ and the intercepts on the v and v/a axes are V and V/K_m, respectively.

All of the plots we have mentioned have commonly been referred to by various personal names, such as the 'Lineweaver – Burk plot', for the double-reciprocal plot. However, these names are often historically inaccurate and misleading, and they are always uninformative, and for that reason we prefer descriptive names for plots.

The last method of plotting the Michaelis – Menten equation that is in current use is conceptually rather different from the others we have described, as it requires each observation to be plotted as a straight line and the parameter values

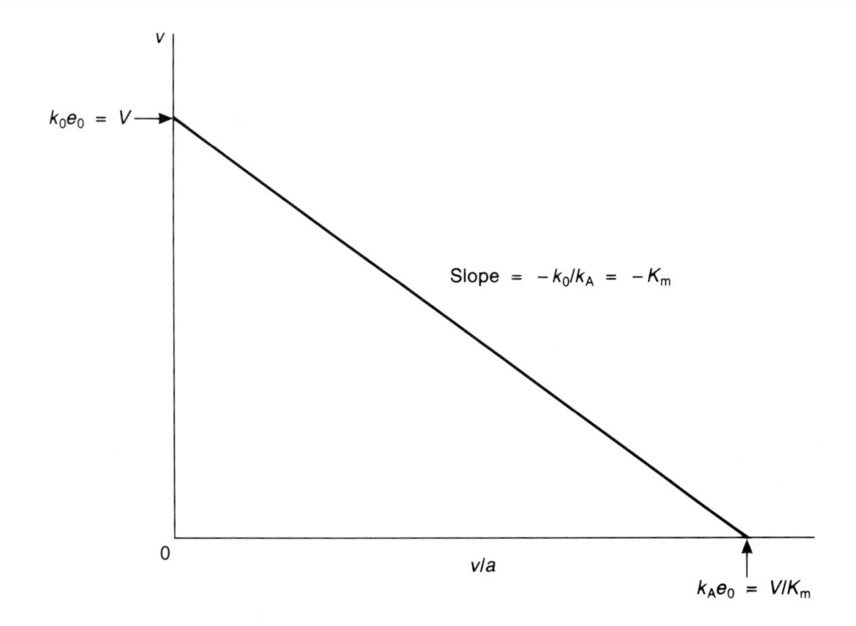

Figure 1.4. The plot of v against v/a.

appear as a point rather than as the slope and intercept of a line. This is the *direct linear plot* and its equation is just a rearrangement of Equation 1.23:

$$V = v + (v/a)\,K_m \tag{1.24}$$

If V and K_m are treated as variables and v and a as constants, this equation defines a straight line with intercepts v on the V axis and $-a$ on the K_m axis. This line relates all possible pairs of (K_m,V) values that exactly satisfy an observation of rate v at substrate concentration a. A second line drawn in the same way for a second observation relates all parameter values that satisfy this second observation. Only one point is on both lines, their point of intersection, and its coordinates define the unique pair of parameter values that satisfy both observations exactly.

In the absence of experimental error, we should expect that n such lines for a set of n observations would all intersect at a unique point, the coordinates of which would give the values of K_m and V. A real experiment, however, is subject to error, and gives a family of intersection points, as illustrated in *Figure 1.5*. It is nonetheless easy to estimate the parameter values because each intersection point provides one estimate of K_m and one estimate of V. These can be marked on the axes and the best estimate of each parameter can be taken as the *median* (middle) value of the set. One takes the median rather than the mean because it is easier, requiring only counting rather than calculation, and

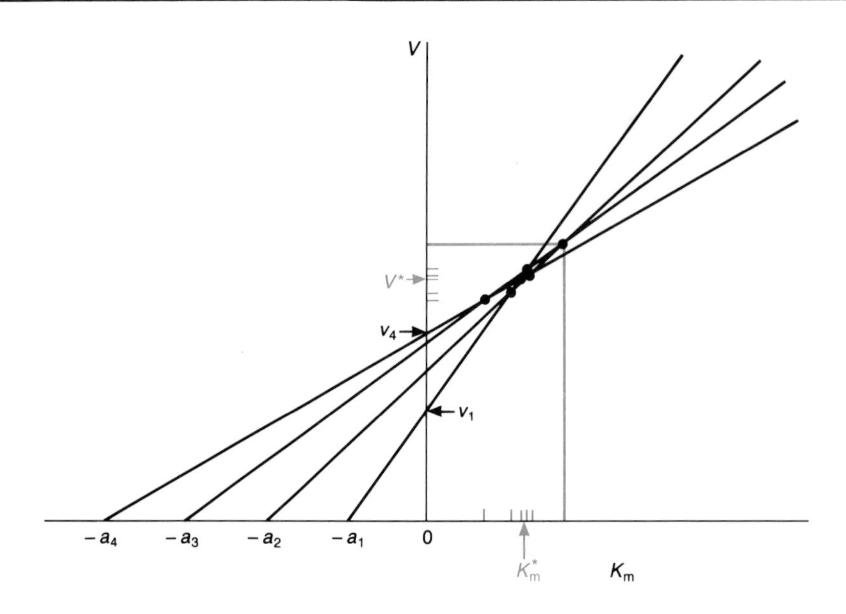

Figure 1.5. The direct linear plot. Each observation is represented not by a point, as in most kinds of plot, but by a straight line, with intercepts $-a$ and v on the abscissa and ordinate, respectively. Each point of intersection provides an estimate of K_m and an estimate of V: this is illustrated in colour for the pair of lines for observations 2 and 3; for the other five intersections the estimates are marked on the axes but complete lines are not shown. The best estimate of K_m is taken as the median (middle) of the ordered set; this is marked on the plot as K_m^*, and the best estimate V^* of V is obtained similarly.

also because it is much safer: some pairs of lines may be nearly parallel and so some of the intersection points may be very far from the correct values; such wildly inaccurate estimates have a severely deleterious effect on the calculation of a mean, but hardly any effect on the median.

Just as there are three ways of plotting the Michaelis–Menten equation as a straight line, there are also three related versions of the direct linear plot. The form shown in *Figure 1.5* is the most familiar, and perhaps the easiest to use, but the alternative (4) shown in *Figure 1.6* has some advantages, of which the simplest is that it provides better-defined intersection points, which all occur in the first quadrant or very close to it.

The direct linear plot is especially valuable for use in the laboratory during the course of an experiment, because it can be plotted as the experiment proceeds and yields parameter estimates without the need for any calculation. It also has some desirable statistical properties (4), which we shall not discuss here beyond mentioning that it does not require weighting and that it is less sensitive to individual erroneous readings than the least-squares estimates (see Section 6).

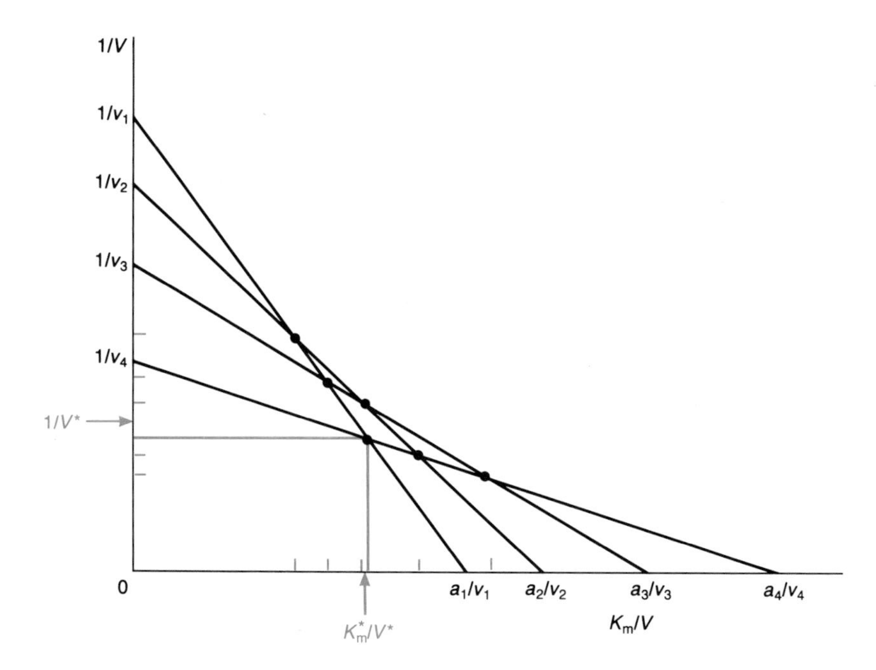

Figure 1.6. Alternative form of the direct linear plot. Here each observation is represented by a line making intercepts of a/v and $1/v$ on the abscissa and ordinate axes, respectively. In other respects the plot is as in *Figure 1.5*.

However, for the final display of results in published work, it tends to present an untidy and crowded appearance, and makes it difficult to display several experiments on one graph; for such purposes the plot of a/v against a is more suitable.

6. Statistical calculation of K_m and V

Many modern electronic calculators include built-in linear-regression functions, so that at the press of a button one can obtain least-squares estimates of the slope and intercepts of a straight line. These functions do not usually make any provision for weighting the individual observations, and although they may remove a little of the subjectivity of estimating parameter values from graphs, they do not avoid any of the inherent statistical problems of the straight-line forms of the Michaelis–Menten equation. An unweighted least-squares fit to a double-reciprocal plot is not appreciably better than drawing a line by eye! This is not the place to enter into a detailed discussion of the statistical aspects of graphs, and we shall therefore say only that in enzyme kinetics *it is virtually never appropriate to calculate an unweighted fit to a straight line.*

Without further discussion, we simply provide formulae that will give reasonable values of the Michaelis–Menten parameters in the presence of the sort of experimental error that we believe to be common in enzyme kinetic measurements. Specifically, they are derived by assuming that each rate is subject to about the same proportion of random error, that is, that the standard error of v is proportional to v:

$$K_m = \frac{\Sigma v^2 \Sigma v/a - \Sigma v^2/a \Sigma v}{\Sigma v^2/a^2 \Sigma v - \Sigma v^2/a \Sigma v/a} \tag{1.25}$$

$$V = \frac{\Sigma v^2/a^2 \Sigma v^2 - (\Sigma v^2/a)^2}{\Sigma v^2/a^2 \Sigma v - \Sigma v^2/a \Sigma v/a} \tag{1.26}$$

In these formulae each summation is over the whole set of data, for example $\Sigma v^2/a^2$ means the sum of all the values of v^2/a^2.

7. Reversible reactions

All chemical reactions are reversible in principle and many biochemical reactions have equilibrium constants that make them fairly easily reversible in practice. However, one can usually 'force' a reaction to be irreversible for practical purposes by measuring initial rates with a zero concentration of at least one product. It is this fact that allows one to ignore the back conversion of E and P into EA in Equation 1.1, not any assumption about the magnitude of the rate constant. Nonetheless, sometimes one needs to consider the rate of a reaction under conditions where the reverse reaction can occur at a significant rate. In this case, the two-step mechanism of Equation 1.7 must be replaced by one in which both steps are reversible:

$$\underset{e_0 - x}{E} + \underset{a}{A} \underset{k_{-1}}{\overset{k_1}{\rightleftharpoons}} \underset{x}{EA} \underset{k_{-2}}{\overset{k_2}{\rightleftharpoons}} E + \underset{p}{P} \tag{1.27}$$

and the expression for the rate of change of concentration of the intermediate EA (Equation 1.8) must be modified by the inclusion of a fourth term representing the rate at which it is being formed from products:

$$dx/dt = k_1 (e_0 - x)a - k_{-1}x - k_2 x + k_{-2} (e_0 - x)p \tag{1.28}$$

Proceeding as in Section 2, the rate of the reaction is:

$$v = \frac{k_1 k_2 e_0 a - k_{-1} k_{-2} e_0 p}{k_{-1} + k_2 + k_1 a + k_{-2} p} \tag{1.29}$$

which may be written as follows, corresponding to Equation 1.12:

$$v = \frac{k_0 e_0 a - k_{-0} K_{mA} e_0 p / K_{mP}}{K_{mA}(1 + p/K_{mP}) + a} \tag{1.30}$$

where k_0, which is equal to k_2, and K_{mA}, equal to $(k_{-1} + k_2)/k_1$, are the catalytic constant and Michaelis constant for the forward direction, defined as in Section 2 (but with an additional subscript A in K_{mA} to distinguish it from K_{mP}), and k_{-0}, which is equal to k_{-1}, and K_{mP}, equal to $(k_{-1} + k_2)/k_{-2}$, are the corresponding parameters for the reverse reaction.

If we put p equal to zero, Equation 1.30 simplifies to Equation 1.12. This is hardly surprising, but the point is still worth making explicitly in order to emphasize that *the use of the irreversible equation implies no assumption about the value of k_{-2}*; it is sufficient for the concentration of product to be zero. Alternatively, if we put a equal to zero then we obtain:

$$v = \frac{-k_{-0} e_0 p}{K_{mP} + p} \tag{1.31}$$

the Michaelis–Menten equation for the reverse reaction. Here the minus sign just reflects the way in which v is defined as the rate of *increase* of p, so that if P is being converted into A this rate of increase is negative.

The reversible equation provides the first clear example of the advantage of regarding the specificity constant as a parameter in its own right: writing k_A equal to k_0/K_{mA} and k_P equal to k_{-0}/K_{mP}, Equation 1.30 assumes the form

$$v = \frac{k_A e_0 a - k_P e_0 p}{1 + a/K_{mA} + p/K_{mP}} \tag{1.32}$$

It can be seen that Equation 1.32 has a much tidier appearance than Equation 1.30, and it allows a simple way of understanding the relationship between the Michaelis–Menten parameters and the *equilibrium constant* of the reaction. If a reaction is at thermodynamic equilibrium its rate is zero, and it is obvious from Equation 1.32 that when v is equal to zero then p/a is equal to K_{eqm}, the equilibrium constant and so k_A/k_P must be equal to K_{eqm}. Given that an enzyme, in common with any other catalyst, cannot alter an equilibrium constant, it follows that the specificity constants for the forward and reverse directions of a reaction are not independent of one another. Once one has been fixed (for example by evolutionary pressure on the enzyme), the other is automatically fixed also by thermodynamic considerations.

Written in terms of k_0 and K_m, or in terms of V and K_m, this relationship between kinetic parameters and the equilibrium constant is known as the *Haldane relationship*:

$$K_{eqm} = k_A/k_P = \frac{k_0 K_{mP}}{k_{-0} K_{mA}} = \frac{V_+ K_{mP}}{V_- K_{mA}} \tag{1.33}$$

where V_+, which is equal to k_0e_0, and V_-, equal to $k_{-0}e_0$, are the limiting rates of the forward and reverse reactions, respectively.

8. Experimental investigation of fast reactions

One of the most notable features of enzyme-catalysed reactions in the cell is that they are very fast, often several orders of magnitude faster than the corresponding uncatalysed reactions. Even the interconversion of bicarbonate and dissolved carbon dioxide, quite a rapid reaction by everyday standards with only water as catalyst (reaching equilibrium in a few seconds), is accelerated in the lungs by the enzyme carbonic anhydrase to the point where it has a half-time of the order of microseconds. The origin of these very high rates is the reason for much of our interest in enzymes as catalysts, but they do not make them easy to study, and nearly all investigations of enzymes are still carried out under steady-state conditions, when the reaction is artificially made very slow by the use of enzyme concentrations much lower than those that exist in the cell.

Most of this book will be concerned with enzymes in the steady state, as the study of fast reactions is usually considered a more advanced and specialized topic. Nonetheless, any student of enzymes ought to be aware that techniques exist for studying enzyme-catalysed reactions under conditions where they occur at a very fast rate, and should realize that additional information becomes available that cannot be derived from steady-state measurements.

The principal methods used for studying fast reactions may conveniently be classified into (i) flow methods, including rapid-mixing and quenching methods, and (ii) perturbation methods. These are not necessarily distinct, as one can use both kinds of method simultaneously, but in an introductory account it is helpful to keep them separate.

8.1 Flow methods

The earliest flow measurements (5) involved a continuous flow of protein (haemoglobin) and substrate (oxygenated water) through a mixing chamber into a long tube, so that observation of the mixture at a particular distance from the point of mixing involved observation of a continuously changing population of molecules, but ones with a fixed 'age' from the instant of mixing. This allowed reactions with a half-time of milliseconds to be studied with equipment requiring several seconds for each measurement, and illustrates that one should not allow the fact that a problem is obviously impossible to solve to deter one from trying to solve it! This was effective, but required very large amounts of protein, and could not be applied to most enzymes. More recently, the standard technique has been the stopped-flow method (6), in which the enzyme and substrate are rapidly pumped through a mixing chamber into a stopped syringe that stops the flow almost instantaneously, and the reaction mixture is observed with rapidly responding automatic equipment. Because of the mechanical violence implicit in this type of experiment there were a number of engineering problems that needed to be solved to produce reliable equipment, but these have now largely been solved, so that commercial equipment is now generally available with a 'dead time', or time between mixing and observation, of less than a millisecond.

The stopped-flow technique requires that the chemical reactions that occur on mixing enzyme and substrate result in spectroscopically observable changes, or a release or consumption of protons that can be coupled to an observable change in the absorbance of an indicator. This may not always be possible or convenient, and then one should use the quenched-flow method, which combines rapid mixing with rapid 'quenching' of the reaction. In this case, after mixing, the reagents pass not into an observation cell but into a vessel that causes the enzyme reaction to stop very rapidly, either because of chemical destruction by trichloroacetic acid or other suitable denaturing reagent, or by being brought to a very low temperature with liquid nitrogen. The quenched solution can then be subjected to chemical analysis on a more leisurely time scale.

To illustrate the sort of information that can be derived from a stopped-flow experiment, consider an enzyme that catalyses a reaction of two substrates, A and B. As we shall discuss in Chapter 3, steady-state methods exist for investigating whether the mechanism requires the enzyme to react first with A, then with B, or vice versa, or whether it can react in either order, but we can obtain this information rather easily (and in a way that is more direct and easier to understand), in a stopped-flow experiment. Suppose that the enzyme cannot react with B until it has bound A: in this case, if enzyme and A are pre-mixed before the start of the stopped-flow experiment, and B is added in the experiment, reaction with B can start immediately, so there will be no lag while A binds to the enzyme; if, however, the enzyme is pre-mixed with B, and A is added in the stopped-flow experiment, there must be an initial lag before any reaction with B is observed. Thus, although the steady-state rates will be identical in the two cases if the concentrations of enzyme and substrates after mixing are the same, the *pre-steady-state phase*, that is the approach of the steady state, will be different depending on which substrate is added first to the enzyme.

8.2 Perturbation methods

Because of the mechanical nature of rapid-mixing and stopping, the stopped-flow method cannot be used to measure processes that occur in less than about 0.5 ms, and it is unlikely that major improvements on this will become possible. For observing very rapid processes, therefore, a different approach is needed, provided by the general class of techniques known as perturbation methods. The best known of these is the *temperature-jump* method (7), in which a reaction mixture at equilibrium or in a steady state is subjected to an electrical discharge that causes the temperature to rise by about 10°C in about 1 μs. In general the equilibrium constant (or steady-state rate equation) will not be the same at the two temperatures, and consequently the reaction mixture will no longer be at equilibrium or steady state at the higher temperature. By observing the 'relaxation' of the reaction mixture to its new equilibrium or steady state, therefore, one can observe processes that occur on a time scale of microseconds.

9. Further reading

Boyde,T.R.C. (1980) *Foundation Stones of Biochemistry*. Voile et Aviron, Hong Kong. [Valuable source of information about the early development of enzymology, and reproduces (in English!) several classic papers.]

Cornish-Bowden,A. (1979) *Fundamentals of Enzyme Kinetics*. Butterworths, Boston. (Contains additional detail about most subjects treated in Chapters 1–5 of this book.)

Cornish-Bowden,A. (1981) *Basic Mathematics for Biochemists*. Chapman and Hall, London. (The mathematics needed for steady-state enzyme kinetics is mainly simply algebra given an appearance of complexity by a profusion of subscripted symbols. Readers with little mathematical background may find a simple review helpful.)

Fersht,A. (1985) *Enzyme Structure and Mechanism*. Freeman, New York (2nd edn). (Excellent account of many subjects treated in this book, especially fast reactions.)

Wharton,C.W. and Eisenthal,R. (1981) *Molecular Enzymology*. Blackie, Glasgow. (Especially relevant to Chapter 6, but also covers elementary material.)

10. References

1. Michaelis,L. and Menten,M.L. (1913) *Biochem. Z.*, **49**, 333.
2. Van Slyke,D.D. and Cullen,G.E. (1914) *J. Biol. Chem.*, **19**, 141.
3. Briggs,G.E. and Haldane,J.B.S. (1925) *Biochem. J.*, **19**, 338.
4. Cornish-Bowden,A. and Eisenthal,R. (1978) *Biochim. Biophys. Acta,* **523**, 268.
5. Hartridge,H. and Roughton,F.J.W. (1923) *Proc. R. Soc. Lond., Series A,* **104**, 376 and 395.
6. Gibson,Q.H. and Milnes,L. (1964) *Biochem. J.,* **91**, 161.
7. Eigen,M. (1954) *Discuss. Faraday Soc.*, **17**, 194.

2

Derivation of steady-state rate equations

1. General considerations

Section 2 of Chapter 1 described the derivation of the Michaelis–Menten equation with the steady-state assumption: Equation 1.8 was written down to express how the concentration of intermediate changed with time, and the rate was then set to zero to give Equation 1.9, which could be rearranged to provide an expression for the concentration of intermediate in the steady state. In principle, exactly the same procedure can be applied to any enzyme mechanism; but it rapidly becomes hopelessly unmanageable: an expression of the type represented by Equation 1.8 must be written down for *each* intermediate in the mechanism, and when each is defined as zero one has a set of simultaneous equations that must be solved to provide the steady-state concentrations. For example, if there are three intermediates there is a set of three simultaneous equations to be solved. It is obvious that this will be extremely laborious and prone to error, even for some of the simpler mechanisms that must be considered. Moreover, much of the labour turns out to be wasted if one approaches the problem in this way, because nearly all of the terms that one has to derive cancel from the final expression.

In this chapter we shall discuss a different way of deriving rate equations termed the method of King and Altman (1). This has the noteworthy characteristic that it is very much easier to apply than it is to understand. For this reason, although in general we believe that it is unwise for anyone to make routine use of a method without understanding why it works, we shall simply describe the method without attempting to demonstrate that it is valid.

We should also emphasize that much of the value of the method of King and Altman to the modern enzymologist lies in the insight that it provides into the kinetic properties of mechanisms, rather than as a tool for actually deriving rate equations. Rate equations for all of the more common enzyme mechanisms are already available from textbooks or from research papers, and those for more complex mechanisms are often too complicated to be obtained without the use of a computer. Nonetheless, with experience of the method of King and Altman it is relatively easy to judge the kinetic characteristics of a mechanism simply by looking at its expression on paper. For this reason it is very useful to gain practice with deriving the rate equations for simple cases even though the results may be well known.

2. Ternary-complex mechanism with dead-end inhibition

Let us consider a mechanism for a reaction of two substrates, A and B, which are required to bind the enzyme E in that order with *on* rate constants of k_1 and k_2, respectively, and *off* rate constants of k_{-1} and k_{-2}, respectively. They generate a ternary complex EAB that releases the products, first P, with a rate constant of k_3, and then Q, with a rate constant of k_4. Suppose, moreover, that the reaction is considered in the presence of an inhibitor I that acts as a non-reactive analogue of B, that is it binds to EA with *on* and *off* rate constants k_i and k_{-i}, respectively, to produce a complex EAI that is incapable of reaction. This is called a *ternary-complex mechanism with dead-end inhibition*, but here we are just concerned with it as an example of a more complex mechanism than the ones we considered in Chapter 1.

The steady-state rate equation for this mechanism is as follows:

$$v = \frac{k_1 k_2 k_3 k_4 e_0 ab}{k_{-1}(k_{-2} + k_3)k_4 + k_1(k_{-2} + k_3)k_4 a(1 + k_i i/k_{-i}) + k_2 k_3 k_4 b + k_1 k_2(k_2 + k_4)ab} \tag{2.1}$$

As this rate expression contains one product of rate constants in the numerator and nine in the denominator (if the parentheses are multiplied out), whereas Equation 1.11 contained one in the numerator and three in the denominator, one might hope that Equation 2.1 would be only around three times as laborious to derive. In reality, however, it would take perhaps 50 times as much effort to derive Equation 2.1 if we used the same approach as we used for Equation 1.10. As there are four intermediates in the mechanism, EA, EAI, EAB and EQ, there would be four rate equations to be set to zero and four simultaneous equations to be solved!

Let us examine Equation 2.1 more closely, with the terms multiplied out and arranged in a way, the logic of which will become apparent, as shown in *Figure 2.1*. Above each term in the equation there is a pattern of individual steps in colour corresponding to the rate constants in the term. For example, k_{-1} is the rate constant for the conversion of EA to E and A, and whenever k_{-1} appears in the equation the reverse reaction of the formation of E from EA appears in the pattern above it. Whenever one has a second-order rate constant such as k_i, the rate constant for the combination of EA and I to give EAI, it occurs as a product with the appropriate reactant concentration, that is as a *pseudo-first-order rate constant*; so we find $k_1 a$, $k_2 b$ and $k_i i$ in the rate equation, but never

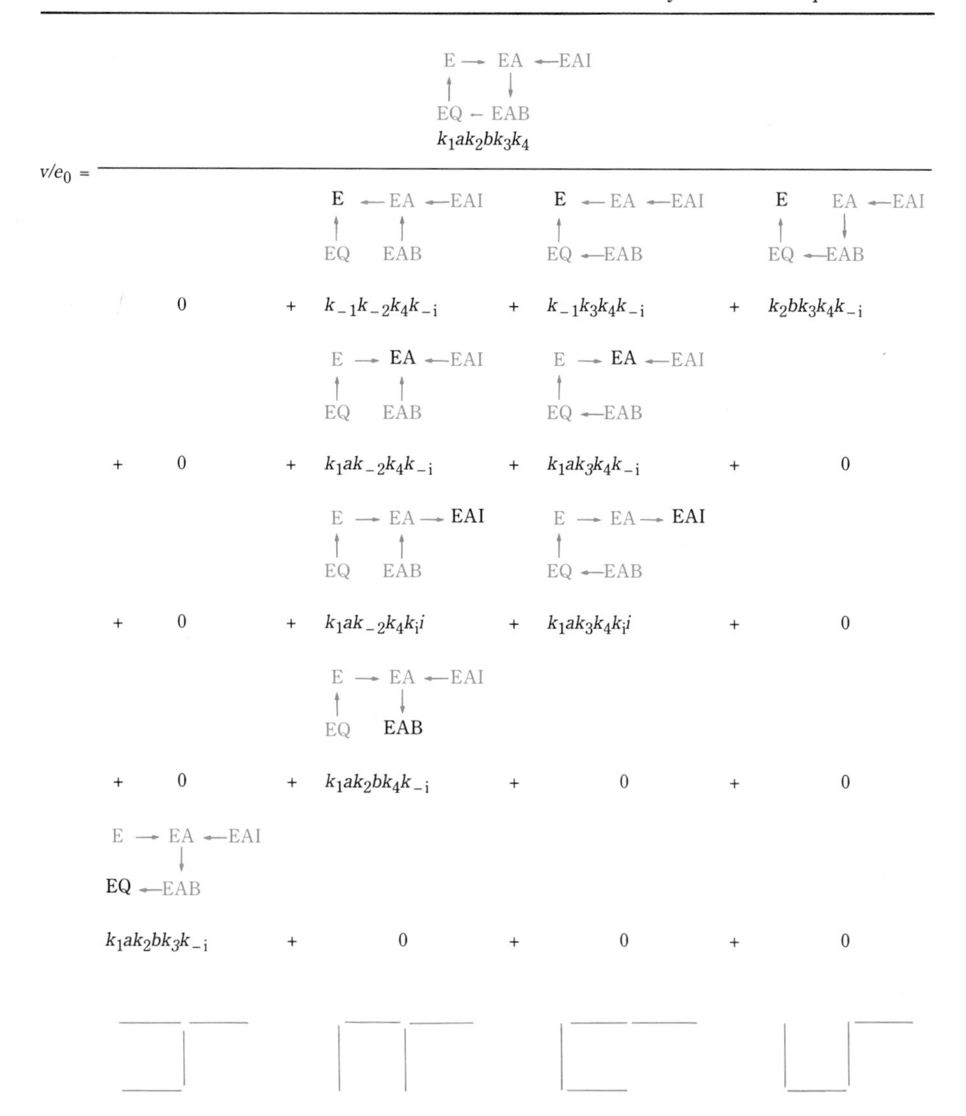

Figure 2.1. Relationship between a mechanism and the rate equation derived from it. See the text for an explanation.

k_1, k_2 or k_i alone. Again the corresponding conversion of EA to EAI, in the case of $k_i i$, appears in the pattern above.

In the first line of the denominator the arrows for every pattern lead to E (not highlighted in *Figure 2.1*) regardless of where one starts. In the other lines the arrows lead to whichever enzyme form is shown in black. Below the equation the patterns are shown in skeleton form without arrow-heads. Further study should show that *all possible patterns of arrows* that connect all of the enzyme forms and terminate at a unique point without any cycles appear in the denominator. The many zeros in the denominator of *Figure 2.1* correspond to patterns that cannot terminate at the required enzyme forms because of irreversible steps. For example, the left-hand pattern contains the irreversible step of EQ being produced from EAB as its only connection to EQ: consequently it can provide a pattern leading to EQ but not to any other enzyme form. If the reaction were considered in the presence of the product (P) so that this step was no longer irreversible, all of the zeros in the left-hand column would need to be replaced by products of rate constants.

The numerator of the rate expression is similar in type, but now the characteristic is that each pattern contains a complete cycle of arrows that accomplishes the chemical transformation A + B $\longrightarrow$ P + Q, with any enzyme form outside the cycle, EAI in this example, connected to it by arrows leading into the cycle. In this example there is only one pattern with these characteristics, and hence only one term in the numerator. If the reaction were considered in the presence of both products, so that there were no irreversible steps, the reverse cycle (E $\longleftarrow$ EA $\longleftarrow$ EAB $\longleftarrow$ EQ $\longleftarrow$ E) would also need to be considered and, as this accomplishes the *reverse* transformation P + Q $\longrightarrow$ A + B, it would be included with a minus sign.

3. The method of King and Altman

In the previous section we started with a rate equation and saw how the terms in it were related to the steps in the corresponding mechanism. The method of King and Altman is simply a formal procedure for reversing this process, starting with a mechanism and generating a rate equation.

The first step is to draw the complete mechanism as a reaction scheme, preferably closed so that no enzyme form occurs more than once. For the mechanism considered in Section 2, this will give a scheme similar to the one drawn above the numerator in *Figure 2.1*, except that the reversible reactions must be shown as reversible. Reactants are *not* included as reactants in this scheme. Instead, when the arrows are labelled with their first-order rate constants, second-order steps are labelled with products that include the appropriate concentrations. Thus, the reaction of E and A to form EA is written as E being converted to EA and labelled with its pseudo-first-order rate constant k_1a, not its second-order rate constant k_1.

We now find all possible patterns that (i) consist of lines corresponding to steps in the mechanism; (ii) connect every enzyme form; and (iii) contain no closed loops. In the example, there are four such patterns, which are drawn in the bottom line of *Figure 2.1*, below the equation. For each enzyme form in turn we now draw (or imagine) arrow-heads on each pattern in such a way that the arrows lead to the enzyme form in question. If the presence of irreversible steps makes this impossible we write zero underneath the pattern; for example, we cannot draw arrow-heads on the left-hand pattern so that it terminates at E. Otherwise we write down the product of first-order rate constants corresponding to the arrows. When this has been done for all of the patterns and all of the enzyme forms, the sum of all the products obtained is the denominator of the expression for v/e_0.

To generate the numerator of the rate expression, we find patterns according to the same rules, (i) and (ii), but a modified rule (iii) that asks for patterns that contain one closed loop capable of accomplishing the chemical transformation, with any enzyme forms outside this loop connected to it by arrows leading into it. Again, we find all possible patterns of this kind, and write down the product of rate constants for each, prefixing it by a negative sign if it accomplishes the chemical transformation in reverse. The sum of such terms is the numerator (2).

The rate equation is complete at this stage, but as an optional final step we will usually wish to tidy it up, so that concentrations are separated from rate constants and terms of a similar kind are collected together. We may, for example, prefer to write the equation of *Figure 2.1* as it appears in Equation 2.1, but this is just a cosmetic exercise.

4. Refinements

The primary objective in applying the method of King and Altman is, of course, to generate a rate equation, but it is worth noting that it provides the *distribution equations* for virtually no extra effort. The proportion of the total enzyme that exists as any particular enzyme form is given by the sum of products for patterns terminating at that form divided by the sum of all the products for all the enzyme forms. For example:

$$[\mathrm{EAI}]/e_0 = k_1(k_2 + k_3)k_4k_i ai/(\text{sum of all nine denominator terms}) \qquad (2.2)$$

The method of King and Altman also provides an easy way to see where all of the terms in a complicated rate equation have come from, and thus to check that they are correct. Moreover, one can easily determine if terms with a particular combination of concentrations can occur in a rate equation without having to derive it. For example, Equation 2.1 contains the product ai but not bi or i without a, showing that I acts as an *uncompetitive* inhibitor with respect to A but as a *competitive* inhibitor with respect to B (cf. Chapter 4, Section 1). With experience of the method of King and Altman this can be deduced just by looking at the mechanism: it is easy to find patterns that generate terms in ai, but impossible to find ones that generate terms in bi or in i without a.

A last point to note about the example we have considered is that the terms for EAI in *Figure 2.1* are exactly the same as those for EA apart from the factor $k_i i/k_{-i}$, or i/K_i, where K_i, equal to k_{-i}/k_i, is the equilibrium constant for dissociation of I from EAI. This means that the binding of I to EA is at equilibrium in the steady state, so that it is not necessary to include the individual rate constants in the rate equation. By studying the rules of the method of King and Altman carefully one should be able to deduce that this is a general feature of *dead-end reactions* within a mechanism: whenever a mechanism contains a reaction or set of reactions that do not lead back into the mechanism in a cycle, this reaction can be treated as an equilibrium even though the productive part of the mechanism is in a steady state.

5. Further reading

Cornish-Bowden,A. (1976) *Principles of Enzyme Kinetics*. Butterworths, London, p. 34. (As far as we know this is the only textbook or review that explains *why* the method of King and Altman works.)

6. References

1. King,E.L. and Altman,C. (1956) *J. Phys. Chem.,* **60**, 1375.
2. Wong,J.T. and Hanes,C.S. (1962) *Can. J. Biochem. Physiol.,* **40**, 763.

3

Reactions of two substrates

1. Types of enzyme mechanism for reactions of two substrates

The usual introduction to enzyme-catalysed reactions is by way of reactions involving one substrate being converted to one product. However, these reactions, catalysed by *isomerases* and *mutases*, are actually quite rare in biochemistry unless we include hydrolytic reactions (which are strictly two-substrate reactions in which the second substrate is water). In reality, the most common type of reaction has two substrates and two products, and is often called a *group-transfer reaction* because it is a reaction in which a group is transferred by the enzyme from a donor to an acceptor.

In an isomerization reaction it is hard to conceive of a sequence of events that does not involve binding of the substrate to the enzyme, followed by one or more chemical reactions within the enzyme – substrate complex, followed by release of product. For group-transfer reactions there are more possibilities to be considered and a large part of the aim of steady-state kinetic investigations is to distinguish between them. For example, does the reaction involve transfer of the group from the donor to the enzyme, followed by a second transfer from enzyme to acceptor, or does the transfer occur in a single step while both donor and acceptor are in the active site of the enzyme? In the former case, variously known as a *double-displacement, substituted-enzyme* or *ping pong* mechanism, one can in principle expect the first product to be released from the enzyme before the second substrate is bound. In the second type of mechanism, known as a *single-displacement, ternary-complex* or *sequential* mechanism, the reaction must pass through a state (the 'ternary complex') with both substrates bound simultaneously to the enzyme. Actually the correspondence between double-displacement and substituted-enzyme mechanisms and between single-displacement and ternary-complex mechanisms is not absolute, because one can imagine that although two displacements occur, the first product might remain bound to the enzyme until after the departure of the second substrate. Indeed, Spector

(1) has made a strong case for believing that single-displacement reactions do not exist, even though there are undoubtedly some reactions that show the kinetics expected for ternary-complex mechanisms. He maintains that when there is stereochemical evidence for an odd number of displacements the number is much more likely to be three than one. Although this remains a minority view, which we shall not pursue here, we shall refrain from using the terms single- and double-displacement to distinguish mechanisms by kinetic means, preferring the less committed terms ternary-complex and substituted-enzyme mechanisms.

A second question that can be asked, mainly in the context of ternary-complex mechanisms, is whether the substrates must bind to the enzyme in a particular order, giving a *compulsory-order* mechanism, or whether either order is possible, giving a *random-order* mechanism? Similar considerations apply to the release of products. In the early development of kinetic methods for studying two-substrate reactions, it was believed that there was a clear distinction to be made here, with NAD-dependent dehydrogenases as examples of compulsory-order enzymes and kinases as examples of random-order enzymes. It now seems more likely that most enzymes show some preference for one route of reaction but that this is not absolute.

2. Michaelis – Menten kinetics in two-substrate reactions

For a reaction obeying a ternary-complex mechanism, the rate in the absence of products is given by an equation of the following form (2 – 4):

$$v = \frac{k_0 e_0 ab}{K_{iA}K_{mB} + K_{mB}a + K_{mA}b + ab} \tag{3.1}$$

in which the *catalytic constant*, k_0, has a similar meaning to the corresponding quantity in the Michaelis – Menten equation (Chapter 1, Section 3); K_{mA} and K_{mB} are the *Michaelis constants* for substrates A and B, respectively; and K_{iA} is (under some simplifying assumptions) the equilibrium dissociation constant of the EA complex. We shall not derive this equation, because its derivation provides an excellent example of the use of the method of King and Altman (Chapter 2); the only comment needed is to mention that although equations are commonly derived in terms of rate constants it is usually convenient to replace the resulting collections of rate constants by less cumbersome symbols: thus k_0 represents $k_3k_4/(k_3 + k_4)$, where k_3 and k_4 are the rate constants for the two product-release steps. It is sometimes convenient to define a term K_{iB} which is equal to $K_{iA}K_{mB}/K_{mA}$ in addition to the constants that appear in Equation 3.1. However, although K_{iA} and K_{iB} have similar algebraic significance their mechanistic interpretations may differ if A and B occupy dissimilar places in the mechanism: thus, if A and B must bind to the enzyme in that order, K_{iA} has a simple interpretation but K_{iB} does not.

For a reaction obeying a substituted-enzyme mechanism the equation is similar, but lacks the constant term in the denominator:

$$v = \frac{k_0 e_0 ab}{K_{mB}a + K_{mA}b + ab} \qquad (3.2)$$

Determining if this constant term has measurable kinetic effects provides a first step in deciding the type of mechanism, and we shall discuss how this can be done. Although Equations 3.1 and 3.2 appear more complex than the Michaelis–Menten equation they are actually of the same form if one considers variation of one substrate concentration at a time. If a is varied and b is held constant, then terms in Equation 3.1 that do not contain a can be treated as constant and collected together:

$$v = \frac{k_0 e_0 b \cdot a}{(K_{iA}K_{mB} + K_{mA}b) + (K_{mB} + b)a} \qquad (3.3)$$

After dividing all the terms by $(K_{mB} + b)$, this can be written as

$$v = \frac{k_0^{app} e_0 a}{K_{mA}^{app} + a} \qquad (3.4)$$

where k_0^{app} and K_{mA}^{app} are the *apparent values* of the catalytic constant and Michaelis constant and are given by the following expressions:

$$k_0^{app} = k_0 b / K_{mB} + b) \qquad (3.5)$$

$$K_{mA}^{app} = (K_{iA}K_{mB} + K_{mA}b)/(K_{mB} + b) \qquad (3.6)$$

As Equation 3.2 differs from Equation 3.1 only in lacking the constant term $K_{iA}K_{mB}$ from the denominator, it follows that it can also be written as Equation 3.4 when a is varied at constant b, and Equation 3.5, which does not contain K_{iA}, applies equally well to this mechanism. The expression for the apparent value of K_{mA} is simpler, however:

$$K_{mA}^{app} = K_{mA}b/(K_{mB} + b) \qquad (3.7)$$

Very similar simplifications of Equations 3.1 and 3.2 result from treating b as variable and a as constant. In either case, therefore, and for either mechanism, if only one concentration is varied, the kinetics can be considered to obey the Michaelis–Menten equation. The only difference from Chapter 1 is that in the one-substrate case the Michaelis–Menten parameters are treated as true constants, whereas in this case they are 'apparent' values, constant only as long as the concentration of the other substrate is constant. However, this difference is hardly fundamental, as all of the 'constants' of enzyme kinetics are constant only so long as the experimental conditions (e.g. pH, temperature, buffer composition, etc.) are constant.

Examination of Equations 3.5 – 7 reveals the practical meanings of the catalytic and Michaelis constants in a reaction of two substrates. Equation 3.5 shows that k_0^{app} approaches k_0 as b approaches infinity. Thus k_0 defines the catalytic capacity of the enzyme when both substrates are at saturation. More generally, for a reaction of more than two substrates, the catalytic constant is the limit approached when all substrates are at saturation. Similarly, as seen in Equations 3.6 and 3.7, the Michaelis constant for any substrate is the limit approached by the measured ('apparent') value when all substrates apart from the one under consideration are at saturation.

3. Kinetic differences between mechanisms

Equations 3.5 – 7 show that for both of the mechanisms we have considered the apparent values of both k_0 and K_m for one substrate are functions of the concentration of the other. In the case of K_m, but not of k_0, one mechanism shows a somewhat more complex type of dependence than the other. This might seem a rather insufficient difference to form the basis of a method for distinguishing between the mechanisms. It becomes clearer, however, if we consider them in relation to the apparent values of k_0 and k_A (= k_0/K_{mA}: see Chapter 1, Section 3) rather than those of k_0 and K_m. We then find that:

$$k_A^{app} = k_A b/(K_{iB} + b) \tag{3.8}$$

for the ternary-complex mechanism (recalling that K_{iB} is equal to $K_{iA}K_{mB}/K_{mA}$), but

$$k_A^{app} = k_A \tag{3.9}$$

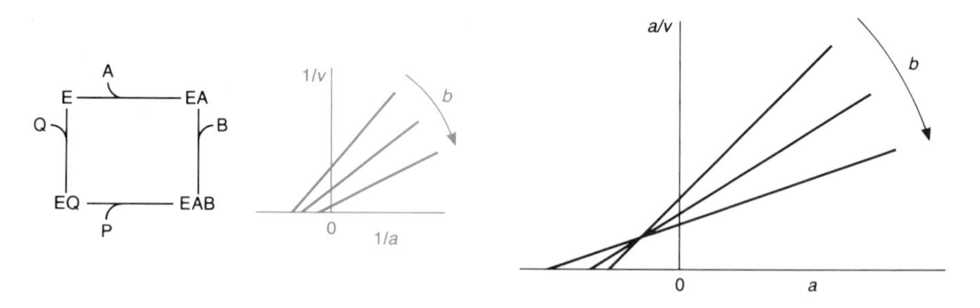

Figure 3.1. Primary plots for a ternary-complex mechanism. If the mechanism shown at top-left is obeyed, plots of a/v against a at several values of b (with b increasing in the direction shown by the arrow) consist of sets of straight lines intersecting in the second (as shown) or third (not shown) quadrant. The appearance of the corresponding double-reciprocal plots (which the student should be able to recognize but is *not* recommended to use) are shown in colour.

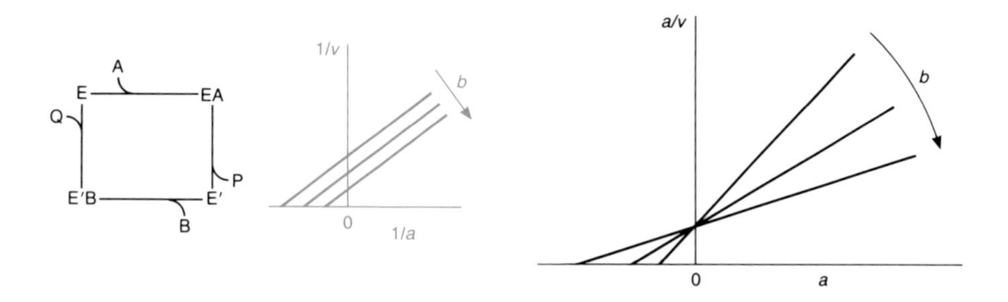

Figure 3.2. Primary plots for the substituted-enzyme mechanism. Other details are as for *Figure 3.1.*

for the substituted-enzyme mechanism. In the former case the apparent value of k_A is a function of b whereas in the latter it is not. This is experimentally convenient because the apparent value of $1/k_A e_0$ is the intercept on the a/v axis in a plot of a/v against a at constant b (cf. Chapter 1, Section 5, Equation 1.22), or, for devotees of the double-reciprocal plot, the slope of a plot of $1/v$ against $1/a$ at constant b (cf. Equation 1.20).

It follows that for plots of a/v against a at several different values of b the intercept on the ordinate decreases as b increases if the enzyme obeys a ternary-complex mechanism (*Figure 3.1*), but all the lines intersect at a common point on the a/v axis if it obeys a substituted-enzyme mechanism (*Figure 3.2*). In both cases the slopes of the lines decrease as b increases.

The types of plot shown in *Figures 3.1* and *3.2* are called *primary plots*, as they are applied directly to the primary data without any preliminary analysis. However, as the apparent parameters are themselves functions of the second substrate concentration b (Equations 3.5–9) we can obtain more information by drawing *secondary plots* of the slopes and ordinate intercepts of the primary plots against $1/b$ (*Figure 3.3*). One can then determine all of the constants in either Equation 3.1 or 3.2. One parameter, K_{iA}, may be obtained directly from the primary plots, as the point at which the lines intersect has an abscissa coordinate of a equal to $-K_{iA}$. The proof of this is not difficult and will be left to the reader, as it provides a useful exercise in the sort of algebra needed for manipulating kinetic equations.

4. Isotope exchange

The sort of method discussed in the previous section provides only a preliminary indication of the type of mechanism in operation, because we have only considered two possible mechanisms out of a wider range that occur in practice, and even within these two cases we have ignored such considerations as

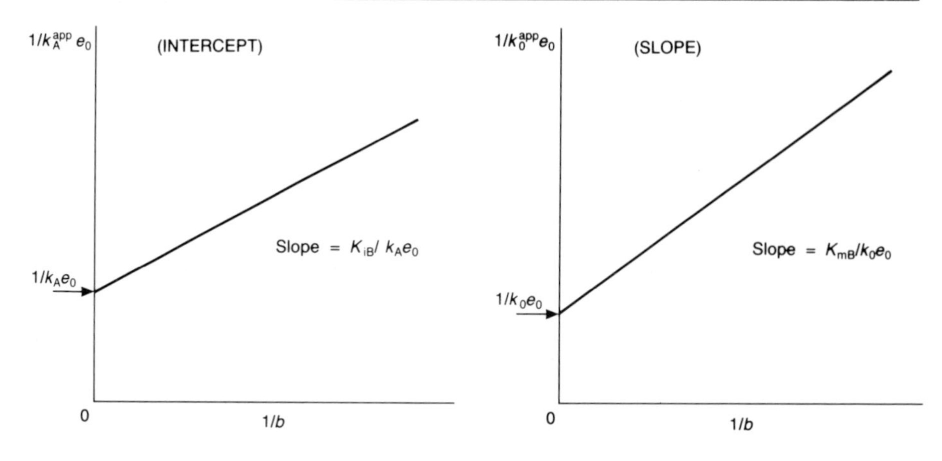

Figure 3.3. Secondary plots for a ternary-complex mechanism. The left-hand and right-hand plots show the dependence on $1/b$ of the ordinate intercepts and slopes, respectively, of the plots of a/v against a shown in *Figure 3.1*. For the substituted-enzyme mechanism (*Figure 3.2*) the plot of primary slopes is similar to the one shown, but the primary intercept shows no dependence on $1/b$.

distinguishing between the two substrates. Moreover, Equation 3.2 is obviously a special case of Equation 3.1, and in practice a ternary-complex mechanism in which K_{iA} is much smaller than K_{mA} will not easily be distinguishable from a substituted-enzyme mechanism. Of more sophisticated ways of distinguishing between mechanisms that need to be considered, we shall defer discussion of *product inhibition* until Chapter 4, after we have discussed inhibition in a more general way. Here we shall give a brief and non-mathematical introduction to *isotope exchange* (5), a very sensitive tool for investigating mechanisms.

A major difficulty with the use of primary plots to distinguish mechanisms is that it tries to use quantitative information to answer a qualitative question: is K_{iA} truly absent from the rate equation, or is it present but small? Isotope exchange allows a yes-or-no type of experiment to be used to answer a yes-or-no question. Consider a reaction mixture consisting of one substrate A and one product P, the other substrate B and the other product Q being absent. If a trace of an enzyme obeying a ternary-complex mechanism is added to this mixture nothing observable will happen, because although binary complexes EA and EP may be formed there is no way to produce a ternary complex in the absence of B and Q, and consequently no possibility of a reaction. If, however, the enzyme obeys a substituted-enzyme mechanism, with A the first substrate and P the first product, the complex EA can release P instead of A, and although there is no *net* reaction in the absence of B and Q there is a continuous interconversion of A and P. In the absence of label this interconversion is not observable, but if one of the components of the mixture is labelled with a radioisotope, the radioactivity of the other component increases until the specific radioactivities are equal.

Isotope exchange thus provides a qualitative difference between ternary-complex and substituted-enzyme mechanisms, a difference in the opposite direction from that seen in primary plots: the mechanism that gives a positive result in isotope exchange is the one that could only be recognized by the lack of a result in the primary plots.

Although ternary-complex mechanisms do not allow isotope exchange to occur in incomplete reaction mixtures, they do allow exchange between particular substrate – product pairs in complete reaction mixtures. Measurement of such exchange can reveal useful information about the details of the mechanism. For a ternary-complex mechanism in which the two substrates, A and B, can only bind to the enzyme in that order, and the two products, P and Q, must be released in that order, the free enzyme E is the only enzyme form capable of reacting with A or Q. Exchange between A and Q can therefore be prevented by setting up conditions in which the free enzyme represents a negligible component of the reaction mixture, for example by extrapolating to infinite concentrations of B and P. Thus, raising B and P to high concentrations can inhibit exchange between A and Q. However, raising A and Q to very high concentrations has no effect on the exchange between B and P and one can only increase this exchange by increasing the availability of EA and EQ, the enzyme forms that react with B and P.

If, however, the substrates in a ternary-complex mechanism can bind in either order, and the products can be released in either order, it is not possible to prevent exchange between a substrate – product pair by raising the concentrations of the other substrate – product pair to saturation.

Isotope-exchange experiments are commonly carried out at chemical equilibrium, that is under conditions where the unlabelled reactants are at equilibrium and only the trace labels are not. This allows much simpler kinetic equations to be used, but it does limit the amount of information that can be obtained. One can also obtain valuable mechanistic information from studies of reaction mixtures in a steady state (6,7). A sensitive method of doing this involves comparing the simultaneous rates of exchange from one reactant to two others, these rates are most conveniently measured by use of doubly labelled reactants. In the forward direction this method provides information about the order of the release of products, and in the reverse direction it provides information about the order of binding of substrates. As the latter is usually of more interest to investigators, we shall suppose that a product P is doubly labelled so that rates of conversion to A and B can be measured simultaneously.

If A and B must bind to the enzyme in that order, so that B is released before A when the reaction is proceeding in reverse, the ratio of the two rates depends on whether an EA molecule that has released B continues to release A or whether instead it reverses direction and regenerates P. This cannot depend in any way on the concentration of A, because A does not react with EA or with any of the enzyme forms on the route from P to A. It does, however, depend on the concentration of B: if this is zero or very small, EA can only release A, and so in this case the flux rates must be equal. However, B can react with EA and convert it back into E, P and Q, and increasing the concentration of B must increase the likelihood that EA, having released B, will return to P without releasing A. This behaviour is independent of any complexities in the release of products; it does not matter whether P and Q are released in a random or a compulsory order, and in the latter case it does not matter which order. The ratio of rates, known as the *flux ratio*, is conveniently determined by comparing the rates of release of radioactive A and B, after correcting for differences in specific radioactivity between the two labels.

5. Relationship of kinetic constants to equilibrium binding constants

The apparent Michaelis constant for any substrate approaches its true value when the concentration of the other substrate is raised to saturation. This follows from Equations 3.6 and 3.7 for either type of mechanism and provides the definition of a Michaelis constant in a reaction with more than one substrate. For a substituted-enzyme mechanism, the apparent Michaelis constant tends to zero when the other substrate concentration is very small, but for a ternary-complex mechanism it approaches a finite value, K_{iA} or K_{iB}.

If the substrates A and B in a ternary-complex mechanism must bind in that order, then lowering the concentration of the *second* substrate, B, towards zero must cause the binding of A to approach equilibrium, because the slower the binding of B becomes the more difficult it becomes to unbalance the equilibrium of the first binding step. Thus K_{iA}, being the experimentally determined Michaelis constant for A when b is very small, should be the same as the true thermodynamic equilibrium constant for dissociation of EA. In principle, measurement of this equilibrium constant in a binding experiment in the absence of B should give the same value as that determined kinetically. Experimentally, this can be done by *equilibrium dialysis*. To do this a buffer containing the enzyme and substrate A is separated by a semi-permeable membrane (permeable to small molecules but not to proteins) from buffer containing A but not the enzyme. Dialysis is continued until equilibrium is achieved, and the concentration of enzyme – substrate complex determined by subtracting the total concentration of A in the enzyme-free buffer (which at equilibrium is assumed to be the free concentration of A on both sides of the membrane) from the total concentration of A in the enzyme-containing buffer.

There should be reasonable agreement between the values of K_{iA} obtained from kinetic and equilibrium measurements if A is the first substrate in a compulsory-order mechanism. It need not be exact because one cannot normally make the measurements under truly comparable conditions. For example, steady-state kinetic experiments normally require very low enzyme concentrations to generate manageably slow rates while binding experiments require very large protein concentrations so that the difference between the total substrate concentrations on the two sides of the membrane will be measurable. Moreover, kinetic experiments, which are completed in a few minutes, are often done at much higher temperatures than equilibrium experiments, where low temperatures are needed to ensure stability of the enzyme during the several hours that may be needed to establish equilibrium across the semi-permeable membrane.

Even if these practical considerations do not cause differences between the K_{iA} values measured in kinetic and thermodynamic experiments, the values may disagree because of *non-specific* binding. If A binds at sites remote from the active site in addition to its normal binding, this may well have no kinetic consequences, but it will contribute to the binding detected in equilibrium measurements and provide a source of disagreement.

Observationally, the apparent Michaelis constant for the second substrate in a compulsory-order mechanism behaves in just the same way as that of the first, approaching a finite limit K_{iB}, but the argument that led us to interpret K_{iA} as an equilibrium constant cannot be applied to K_{iB}. Whether the binding of B to EA approaches equilibrium depends on the relative rate constants for release of B and the rate of conversion to products, and not on the concentration of A. Thus K_{iB} is not an equilibrium constant, and for that reason one may prefer to write it as $K_{iA}K_{mB}/K_{mA}$, more cumbersome but less open to misinterpretation.

Thus, there need not be any agreement between K_{iB} measured kinetically and by equilibrium dialysis. Substrate B need not bind at all in an equilibrium-dialysis experiment, though in more complex forms of the mechanism it may bind to some extent, because of non-specific binding at sites remote from the catalytic site, or because of binding at the catalytic site in a way that prevents subsequent binding of A in a proper mode for catalysis. Any binding constant that results from such effects will not agree with the kinetic parameter K_{iB}.

6. Further reading

Spector,L.B. (1982) *Covalent Catalysis by Enzymes*. Springer-Verlag, Berlin. (Ref. 1 below. Even if one has reservations about the general thesis of this book, it is extremely valuable as a source-book for kinetic studies of enzymes.)

See also the general references listed at the end of Chapter 1.

7. References

1. Spector,L.B. (1982) *Covalent Catalysis by Enzymes*. Springer-Verlag, Berlin.
2. Segal,H.L., Kachmar,J.F. and Boyer,P.D. (1952) *Enzymologia*, **15**, 187.
3. Alberty,R.A. (1953) *J. Am. Chem. Soc.*, **75**, 1928.
4. Dalziel,K. (1957) *Acta Chem. Scand.*, **11**, 1706.
5. Boyer,P.D. (1959) *Arch. Biochem. Biophys.*, **82**, 387.
6. Britton,H.G. (1985) In Tipton,K.F. (ed.), *Protein and Enzyme Biochemistry*. Elsevier, Limerick, p. 1.
7. Cornish-Bowden,A. (1988) *Curr. Top. Cell Reg.*, **30**, in press.

4

Inhibition of enzyme activity

1. Types of inhibition

Substances that cause enzyme-catalysed reactions to proceed more slowly when they are present in the reaction mixture are called *inhibitors* and the phenomenon is called *inhibition*. It is an important aspect of enzymology both because many physiological effects, toxic as well as functional, are consequences of inhibition, and because the study of inhibition in the laboratory often yields useful information about enzyme mechanisms.

Inhibition does not necessarily result from a direct interaction between the enzyme and inhibitor—it may, for example, result from the binding of the substrate to the inhibitor to produce a molecule incapable of acting as a substrate—but the most important kinds of inhibition do result from binding of the inhibitor to the enzyme, and in this book we shall restrict our attention to these. We shall also assume that the inhibition is *reversible* by simple treatments such as lowering the inhibitor concentration by dilution or dialysis. Irreversible inhibition, often called *modification*, and usually the result of a covalent interaction between enzyme and inhibitor, is also important for the mechanistic study of enzymes, but it is somewhat apart from ordinary kinetic investigations and we shall not discuss it here.

Even with these restrictions there are several kinds of inhibition, and we shall further restrict attention to the simplest type of inhibition, known as *linear* inhibition, which is manifested by the appearance of terms proportional to the inhibitor concentration in the denominator of the rate equation. (If terms proportional to the square of the inhibitor concentration appear in the denominator the inhibition is *parabolic*, whereas if linear terms appear in the numerator as well as the denominator it is *hyperbolic*. These are encountered much less frequently than linear inhibition.)

When an enzyme is subject to linear inhibition, the reaction still obeys Michaelis–Menten kinetics (if it did in the absence of inhibitor), but with apparent specificity constants and catalytic constants that vary with the inhibitor concentration. If the inhibitor affects only the apparent specificity constant it is called a *competitive* inhibitor; if it affects only the apparent catalytic constant it is called an *uncompetitive* inhibitor; if it affects both constants it is called a *mixed* inhibitor. These descriptions assume nothing about the underlying mechanism of the inhibition: they are simply *operational* definitions that allow one to describe observations before they can be interpreted. Nonetheless, as we shall see, certain simple mechanisms for inhibition are characterized by particular types of inhibition, and many workers prefer to regard these mechanisms as fundamental to the definitions. This has obvious attractions and for competitive inhibition the simplest mechanism makes good sense in relation to the everyday idea of competition, but it also creates difficulties: it means that we cannot describe the data until we have interpreted them, and it means that mechanisms that produce exactly the same kind of behaviour as the simplest mechanism for competitive inhibition are left without a name. This difficulty is compounded in the case of uncompetitive inhibition, where the 'textbook' mechanism for it is not often encountered with real enzymes.

When Michaelis and his collaborators first studied inhibitors of invertase in the early part of this century, they thought that there were some inhibitors that prevented substrate binding and others that interfered with catalysis but had no effect on binding (1). With more information about enzyme mechanisms it has become evident that it is not very likely for an inhibitor to interact with an enzyme and alter its catalytic potential without having any effect on the binding of the substrate and, if we exclude pH effects in which the proton can be regarded as an inhibitor, nature has not provided us with any examples. Unfortunately, however, most general biochemistry textbooks are written by people more interested in molecular biology than in kinetics who think that nothing has been learned about enzyme kinetics since 1934. They continue to devote space to the discussion of phenomena that do not occur in nature while omitting to mention others that do. Readers who wonder why the term 'non-competitive' does not occur in this book (other than in this sentence) may find the explanation in a careful reading of this paragraph.

2. Competitive inhibition

The commonest kind of inhibition (though not as common as is supposed by experimenters who do not check for any other possibilities) is competitive inhibition. It arises most simply if the inhibitor and substrate *compete* for the same form of the enzyme, so that when either binds the other cannot, as illustrated in *Figure 4.1*, in which the inhibitor reacts in a *dead-end* reaction that does not lead to products; alternatively, the inhibitor may be the product of the reaction and binds to the same enzyme form as the substrate. These are the commonest ways in which competitive inhibition can arise, but they are not the only ones: for example, if a substrate directly displaces a product in a concerted step, one will not be able to write any King – Altman patterns that contain both

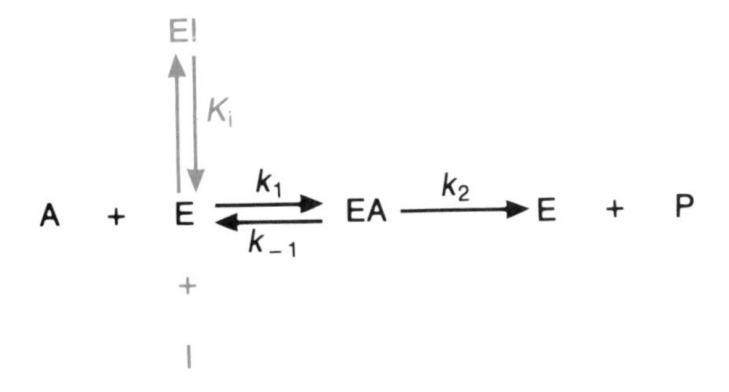

Figure 4.1. Dead-end mechanism for competitive inhibition. The part of the mechanism shown in colour represents the additional part caused by the presence of inhibitor.

concentrations, and the substrate–product pair will act as competitive inhibitors with respect to one another even though they bind to different enzyme forms. We shall discuss competitive inhibition in the context of the dead-end mechanism, but much of what we say applies equally well to the other cases.

Applying the methods of Chapter 2 one may readily show that the rate equation corresponding to *Figure 4.1* is as follows:

$$v = k_0 e_0 a/[K_m(1 + i/K_i) + a] \tag{4.1}$$

where k_0, which is equal to k_2, and K_m, which is equal to $(k_{-1} + k_2)/k_1$, are the catalytic constant and Michaelis constant, respectively, of the uninhibited reaction, and K_i is the dissociation constant of the enzyme–inhibitor complex (EI) commonly known as the *competitive inhibition constant* of the inhibitor, or just the inhibition constant if it is clear what type of inhibition is being considered.

Equation 4.1 is of the form of the Michaelis–Menten equation, with apparent parameters as follows:

$$k_0^{\text{app}} = k_0 \tag{4.2}$$

$$K_m^{\text{app}} = K_m(1 + i/K_i) \tag{4.3}$$

$$k_A^{\text{app}} = k_A/(1 + i/K_i) \tag{4.4}$$

The effect of a competitive inhibitor is therefore to decrease the apparent affinity of the substrate for the enzyme without any effect on the reactivity of the enzyme–substrate complex once formed. As we shall see when discussing the other kinds of inhibition, Equation 4.4 is more fundamental than Equation 4.3, and so we should also note that a competitive inhibitor decreases the apparent value of the specificity constant, k_A (= k_0/K_m).

Remembering that $1/k_0e_0$ is the slope and $1/k_Ae_0$ the ordinate intercept of a plot of a/v against a (Chapter 1, Section 5), we further note that a competitive inhibitor increases the ordinate intercept of such a plot while leaving the slope unchanged: plots drawn at several values of i give parallel lines. (In the double-reciprocal plot the slope and ordinate intercept are reversed, and one gets a family of lines intersecting on the ordinate axis.)

The most direct way to measure the inhibition constant K_i is to determine k_A^{app} at several different values of i and plot $1/k_A^{app}$ against i: Equation 4.4 shows that this gives a straight line with a slope of $1/k_A K_i$ and (more simply) an intercept $-K_i$ on the i axis. Another type of plot, known as a *Dixon plot* (2), is also used, and is perhaps preferable when the number of different inhibitor concentrations is larger than the number of different substrate concentrations. This involves plotting $1/v$ against i at each of several substrate concentrations. It follows from Equation 4.1 in reciprocal form:

$$1/v = (1/k_0 e_0) + (K_m/k_0 e_0 a)(1 + i/K_i) \qquad (4.5)$$

that each plot gives a straight line, though with slope and intercepts that are not immediately informative. At any value of i the value of $1/v$ varies with a unless the factor $(1 + i/K_i)$ in the last term is zero, that is, unless i is equal to $-K_i$. Thus the lines at different a values must intersect at the (extrapolated) point where i is equal to $-K_i$ and $1/v$ to $1/k_0 e_0$.

It is sometimes believed that the appearance of the Dixon plot is diagnostic for competitive inhibition, but this is wrong. Plots with the same general appearance (though with a different interpretation of the $1/v$ value at the point of intersection) are obtained with mixed inhibition, as we shall consider in detail in Section 4.

3. Uncompetitive inhibition

Uncompetitive inhibition is at the opposite extreme from competitive inhibition and it is characterized by the following equation:

$$v = k_0 e_0 a/[K_m + a(1 + i/K_i)] \qquad (4.6)$$

which shows that Michaelis–Menten kinetics is obeyed with an apparent specificity constant independent of inhibitor concentration and an apparent catalytic constant that decreases as the inhibitor concentration increases:

$$k_0^{app} = k_0/(1 + i/K_i) \qquad (4.7)$$

$$K_m^{app} = K_m/(1 + i/K_i) \qquad (4.8)$$

$$k_A^{app} = k_A \qquad (4.9)$$

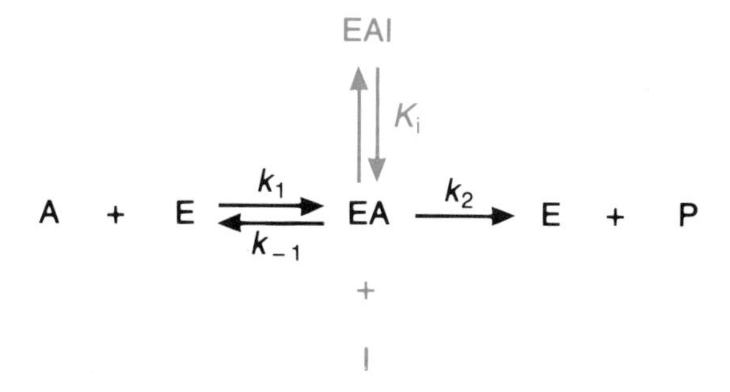

Figure 4.2. Dead-end mechanism for uncompetitive inhibition.

Conceptually the simplest mechanism for uncompetitive inhibition is one in which the inhibitor binds in a dead-end fashion to the enzyme–substrate complex but does not interact with the free enzyme (*Figure 4.2*). The *uncompetitive inhibition constant* K_i is then the dissociation constant of the enzyme–substrate–inhibitor complex EAI for dissociation of inhibitor. This is not a particularly plausible mechanism, because in general it is unlikely that a substance with affinity for the enzyme–substrate complex will have no effect on the free enzyme. Examples of dead-end uncompetitive inhibition are thus uncommon (3), though one of agricultural importance may be found in the herbicide 'Glyphosate' (*N*-phosphonomethylglycine), which owes its potency to its effect as an uncompetitive inhibitor of an enzyme in the shikimate pathway (4).

Uncompetitive inhibition is mainly important as a type of product inhibition, but even then it mainly occurs as a limiting case of mixed inhibition rather than as a phenomenon in its own right. True uncompetitive product inhibition requires the binding of product to be isolated from the binding of substrate by irreversible reactions, for example if the product inhibitor is the second of three products; if the other two products are absent from the reaction mixture its release is flanked by two irreversible product-release steps. (It is useful for testing one's understanding of the method of King and Altman described in Chapter 2 to prove, *without deriving a complete rate equation*, that when substrate- and product-binding steps are isolated from one another by irreversible steps one cannot write down any terms for the rate equation that contain only one of the two concentrations.) This condition can be approached, but only as a limit, if a step is made near-irreversible by a large equilibrium constant or if one or more substrate-binding steps approaches saturation.

The Dixon plot (Section 2) gives parallel lines if one tries to use it with uncompetitive inhibition, and so it does not provide the value of K_i. However, there is a complementary plot that does: if one plots a/v against i there is an intersection point at which i is equal to $-K_i$ in uncompetitive inhibition but this plot gives parallel lines with competitive inhibition (5).

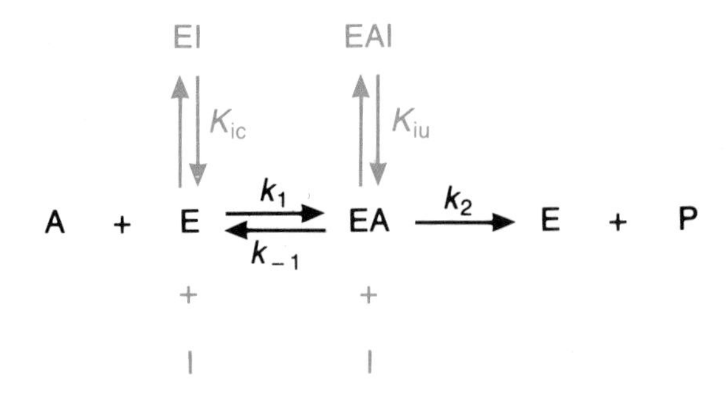

Figure 4.3. Mixed-inhibition. As there are now two inhibitory reactions, with different dissociation constants, different symbols K_{ic} and K_{iu} are needed to differentiate them.

4. Mixed inhibition

If an inhibitor binds to both free enzyme and enzyme–substrate complex, with dissociation constants K_{ic} and K_{iu}, respectively (*Figure 4.3*), both competitive and uncompetitive effects occur simultaneously, and the rate is:

$$v = k_0 e_0 a/[K_m(1 + i/K_{ic}) + a(1 + i/K_{iu})] \quad (4.10)$$

This again accords with Michaelis–Menten kinetics, but now the apparent parameters are as follows:

$$k_0^{app} = k_0/(1 + i/K_{iu}) \quad (4.11)$$

$$K_m^{app} = K_m(1 + i/K_{ic})/(1 + i/K_{iu}) \quad (4.12)$$

$$k_A^{app} = k_A/(1 + i/K_{ic}) \quad (4.13)$$

Not only are Equations 4.11 and 4.13 exactly the same as Equations 4.7 and 4.4, respectively, apart from the inclusion of the additional subscripts to distinguish between the two different inhibition constants in Equation 4.10, but they are also of exactly the same *form* as one another. By contrast, the behaviour of the apparent value of K_m seen in Equations 4.3, 4.8 and 4.12 is irregular and unmemorable. These examples should be studied carefully by anyone who doubts that the fundamental parameters of steady-state enzyme kinetics are k_0 and k_A, and that it is easier to think of K_m as k_0/k_A than to think of k_A as k_0/K_m. In fact, treating K_m as k_0/k_A provides the only easy way of remembering how its apparent value changes in the case of each different kind of inhibition.

Equation 4.10 obviously describes a type of inhibition that contains both competitive and uncompetitive components and so its name of *mixed inhibition* follows naturally. It is the general type of dead-end inhibition when the inhibitor and substrate bind at different sites on the enzyme. Of course, if substrate and inhibitor can bind simultaneously, one might expect the enzyme–substrate–inhibitor complex to show some capacity to form products: this is certainly reasonable, but we shall not consider it in this book because the resulting inhibition is not linear; extra terms appear in the numerator as well as the denominator of the rate equation, and the inhibition is *hyperbolic*. Its most easily observable characteristic is that the rate does not tend to zero as the inhibitor concentration approaches infinity.

Mixed inhibition is also important as the general case of product inhibition, for which linear mixed inhibition should be regarded as the norm, all other types being special or limiting cases.

If one remembers that mixed inhibition is the simultaneous occurrence of competitive and uncompetitive inhibition, it is easy to remember how to recognize it and determine the inhibition constants that characterize it. It is recognized by the fact that the inhibitor affects the apparent values of both k_A and k_0, and its inhibition constants can be measured by plotting the reciprocals of these apparent values against the inhibitor concentration, the former giving the competitive inhibition constant K_{ic}, the latter giving the uncompetitive inhibition constant K_{iu}. The Dixon plot also provides K_{ic} exactly as in competitive inhibition (and the plot has the same appearance as in competitive inhibition), and the plot of a/v against i provides K_{iu} exactly as in uncompetitive inhibition.

There is no reason to expect K_{ic} and K_{iu} to be equal (if the inhibitor eliminates the catalytic potential of the enzyme, why should it have no effect whatever on substrate binding?), and so clearly they require different symbols. One may also use the same pair of symbols for the simple cases described in Sections 2 and 3, but if there is only one inhibition constant to consider most workers feel that the simple symbol K_i is adequate for it. One sometimes sees K_{is} for K_{ic} and K_{ii} for K_{iu}, where the s and i stand for slope and (ordinate) intercept, respectively. However, *we do not recommend this usage*, because the slope and intercept referred to are the slope and intercept of the double-reciprocal plot, the least satisfactory of the common plots of the Michaelis–Menten equation, and the two inhibition constants have meanings, and ought to have symbols, that are independent of the method one may choose to use for measuring them.

5. Competing substrates

The term *specificity* is rather over-worked in enzymology, being often used without any clear definition in mind. An enzyme may be said to be more specific for substrate A than for substrate A′ if it has a higher catalytic constant for A, or a smaller Michaelis constant, or even if it just gives a faster reaction with A than with A′ under some arbitrary set of conditions. A moment's reflection will show that the only physiologically sensible way to define specificity is in relation to the sort of conditions that may occur in the living organism. In these conditions one can only meaningfully say that an enzyme is more specific for

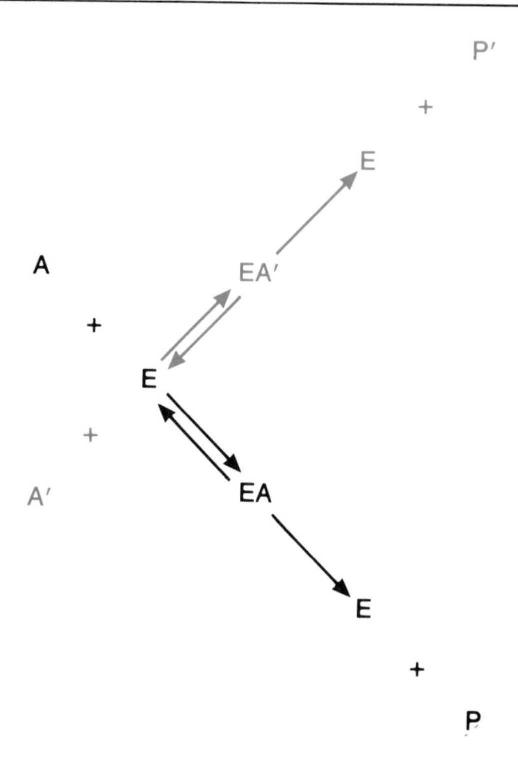

Figure 4.4. Competing substrates. The reaction of A′ shown in colour represents a second reaction that competes with A for the free enzyme.

A if A reacts faster than A′ *when both substrates are available simultaneously in the same solution at the same concentration*. Thus we must consider the kinetics of a mixture of competing substrates for the same enzyme.

An appropriate mechanism is shown in *Figure 4.4*, where the enzyme follows a Michaelis–Menten type of mechanism for either substrate considered in isolation. Defining v and v' as the rates at which A and A′ at concentrations a and a', respectively, are converted into their products, we may use the methods of Chapter 2 to show that these rates are:

$$v = k_0 e_0 a/[K_m(1 + a'/K'_m) + a] \tag{4.14}$$

$$v' = k'_0 e_0 a'/[K'_m(1 + a/K_m) + a'] \tag{4.15}$$

where k_0 and K_m are the catalytic constant and Michaelis constant for the reaction of A in isolation, and k'_0 and K'_m are the corresponding quantities for A′. Both equations are of the same form as Equation 4.1, so that a competing substrate behaves, so far as the other reaction is concerned, like a competitive inhibitor with an inhibition constant equal to its ordinary Michaelis constant. To compare the two rates it is convenient to write the two equations with the same denominator, by dividing all terms in Equation 4.14 by K_m and all terms in Equation 4.15 by K'_m:

$$v = (k_0/K_m)e_0a/(1 + a/K_m + a'/K'_m) \qquad (4.16)$$

$$v' = (k'_0/K'_m)e_0a'/(1 + a/K_m + a'/K'_m) \qquad (4.17)$$

It is then simple to divide one by the other to give the ratio of rates:

$$v/v' = \frac{(k_0/K_m)a}{(k'_0/K'_m)a'} = k_A a/k_{A'}a' \qquad (4.18)$$

from which it is plain that the ability of the enzyme to distinguish between the two substrates, that is its *specificity*, is determined by their concentrations and their ratios of the catalytic constant divided by the Michaelis constant, that is the ratio of the specificity constants. Thus, it is obvious why the name *specificity constant* is appropriate for this parameter.

When we first introduced the specificity constant, in Section 3 of Chapter 1 in the context of Equation 1.13, we saw that for an isolated substrate the specificity constant is the second-order rate constant for the reaction at very low concentrations, that is it defines the rate at these concentrations. We should avoid concluding from this that its meaning as the determinant of specificity is also restricted to low concentrations: Equation 4.18 was derived without any assumption about the magnitudes of a and a', and the specificity constant defines the specificity of an enzyme for a particular substrate at *any* concentration, not necessarily a low one. Equation 4.18 can be generalized to mixtures of more than two competing substrates, so that if substrates A, A′, A″ etc. are present in the same reaction mixture, the rates v, v', v'' etc. are proportional to the appropriate specificity constants multiplied by the corresponding concentrations:

$$v/k_A a = v'/k_{A'}a' = v''/k_{A''}a'' = \ldots \qquad (4.19)$$

6. Product inhibition as a mechanistic probe

We can now resume the discussion of product inhibition as a method for discriminating between mechanisms, deferred from Section 4 of Chapter 3 until the general idea of inhibition had been introduced. Examining Equation 4.10, the equation for linear mixed inhibition, and comparing it with Equations 4.1 and 4.6, the equations for competitive and uncompetitive inhibition, we note that the denominator of the rate expression may in principle contain two inhibitory factors of the form $(1 + i/K_i)$: one that affects the 'constant' part of the denominator, which in the case of Equation 4.10 means K_m, but in general means the part of the denominator that is not affected by a (so it may contain concentrations other than a); and a second that affects the 'variable' part, which is the part that has a as a factor. Moreover, a competitive inhibitor affects only the 'constant' part, an uncompetitive inhibitor affects only the 'variable' part, and a mixed inhibitor affects both.

The full (reversible) rate equation for the compulsory-order ternary-complex mechanism with A binding before B and P released before Q is:

$$v = \frac{e_0(\&ab - \&pq)}{\&1 + \&a + \&b + \&p + \&q + \&ab + \&ap + \&bq + \&pq + \&abp + \&bpq} \quad (4.20)$$

whereas the one for the substituted-enzyme equation is:

$$v = \frac{e_0(\&ab - \&pq)}{\&a + \&b + \&p + \&q + \&ab + \&ap + \&bq + \&pq} \quad (4.21)$$

In both we have adopted an unusual (but we hope comprehensible) symbolism to concentrate on the important features of the equations at the expense of the less important complexities: each '&' in the equations represents some collection of rate constants; for example, $\&a$ in Equation 4.20 is equivalent to $k_1(k_{-2} + k_3)k_4a$, but it has a different (and equally unmemorable) meaning in Equation 4.21. Fortunately for the student of enzyme kinetics, one does not often need to know the detailed definitions of all the terms in a rate equation: to discuss what types of product inhibition are expected the questions are of the form 'is there a term in bp in the rate equation?', not 'what is the coefficient of bp in the rate equation?'.

If we now examine Equations 4.20 and 4.21, the first point is that we normally try to study product inhibition under irreversible conditions, which means that no more than one product is added to the reaction mixture at a time. We can then ignore any terms that contain *both* p and q: $-\&pq$ in both numerators, $\&pq$ in both denominators and $\&bpq$ in Equation 4.20. Looking at what is left, the experimenter needs to decide which substrate – product pair is being considered: which product is present as an inhibitor, and which substrate concentration is to be varied? (The other substrate will, of course, be present, but at a constant concentration.) Suppose that P is the inhibitor and A is the variable substrate: p occurs in Equation 4.20 both in a 'constant' term ($\&p$) and in two 'variable' terms ($\&ap$ and $\&abp$). Thinking of it, therefore, as the inhibitor in Equation 4.10, it is clear that it is in both the constant and variable parts of the denominator, and is thus a mixed inhibitor when A is the variable substrate. By contrast, q appears only in 'constant' terms, $\&q$ and $\&bq$ (remembering that here 'constant' means 'independent of a'), and Q is competitive with respect to A. Similar analysis of Equation 4.20 with B as variable substrate shows that both P and Q are mixed inhibitors with respect to B.

The same analysis can be applied to Equation 4.21, showing that for a substituted-enzyme mechanism P is a mixed inhibitor with respect to A but a competitive inhibitor with respect to B, whereas Q is a competitive inhibitor with respect to A and a mixed inhibitor with respect to B.

These results may seem rather disappointing, and hardly worth the effort of deriving, as the inhibition patterns are the same for the two mechanisms apart from the B – Q combination, and in the compulsory-order ternary-complex mechanism three out of the four substrate – product pairs show mixed inhibition. However, we can greatly emphasize the differences between mechanisms and between substrate – product pairs by considering how the inhibition patterns change when the 'constant' substrate is saturating.

Experimentally we cannot achieve an infinite (truly saturating) concentration, but we can observe the trend as the 'constant' substrate concentration is made large. As b is made larger in Equation 4.20, terms that do not contain b become progressively less important in determining the behaviour. This means that the 'constant' term &p becomes less important by comparison with the 'variable' term &abp, and so the mixed inhibition exhibited by P with respect to A becomes like uncompetitive inhibition as b is increased. In the case of Q, however, the 'constant' terms include &bq, which remains important as b becomes large, and the competitive inhibition of Q with respect to A is maintained. Conversely, when B is the variable substrate, increasing a to saturation does not affect the mixed character of inhibition by P, because ap occurs in both the 'constant' term &ap and the 'variable' term &abp; the mixed inhibition by Q, however, disappears altogether (i.e. Q ceases to be an inhibitor of any kind) as a increases, because there are no terms in aq in the rate equation. Thus of the substrate – product combinations originally giving mixed inhibition, one has maintained its character, one has tended towards uncompetitive inhibition, and the third has disappeared.

When we consider the equation for the substituted-enzyme mechanism in the same sort of way, we find that the competitive inhibition combinations (A – Q and B – P) are maintained when the constant substrate approaches saturation, whereas the mixed inhibition combinations (A – P and B – Q) tend towards no inhibition.

7. Further reading

Fersht,A. (ed.) (1985) *Enzyme Structure and Mechanism*. Freeman, New York, 2nd edn. (Particularly authoritative in relation to enzyme specificity.)

See also the other general references at the end of Chapter 1.

8. References

1. Michaelis,L. and Pechstein,H. (1914) *Biochem. Z.*, **60**, 79.
2. Dixon,M. (1953) *Biochem. J.*, **55**, 170.
3. Cornish-Bowden,A. (1986) *FEBS Lett.*, **203**, 3.
4. Boocock,M.R. and Coggins,J.R. (1983) *FEBS Lett.*, **154**, 127.
5. Cornish-Bowden,A. (1974) *Biochem. J.*, **137**, 143.

5

The pH-dependence of enzyme-catalysed reactions

1. Introductory considerations

Study of the pH-dependence of enzyme-catalysed reactions can provide important information about the nature of the amino acid side chain groups that participate in catalysis. Nonetheless, several serious pitfalls must be avoided if valid conclusions are to be derived. The most important of these concerns the elimination of gross structural effects (denaturation) on the enzyme as the pH is varied and the effect of the protein micro-environment on the pK_a values of the catalytic groups. We shall discuss these sources of ambiguity as well as some others that are more subtle.

It is obvious that catalytic groups must be capable of ionizing within the pH range over which the enzyme has catalytic activity if pH-dependence is to be observed. In principle one can determine the pK_a values of these groups from pH-dependence data and compare these with standards to identify the chemical nature of the catalytic groups.

The side chains of the free amino acids that ionize within the physiological range are given in *Table 5.1*, with some values that have been determined for the same groups in the active centres of enzymes. Quite large perturbations are possible, that is, active-centre groups can have pK_a values substantially different from those of the corresponding amino acids in free solution. These may result from hydrophobicity, hydrogen-bonding, or electrostatic interactions.

A hydrophobic environment tends to destabilize charged species and so will lower the pK_a of a cationic acid, such as the imidazolium ion of histidine, but will raise the pK_a of a neutral acid such as the carboxyl side chain of glutamic acid. The effect of hydrogen-bonding will depend on whether the ionizing species donates or accepts a hydrogen bond. Thus a carboxylate anion can be stabilized by forming a stronger bond with a donor than can the neutral protonated form. Electrostatic interactions, which are strong only in a non-polar environment such as the interior of a protein, can result in stabilization of a charged species when it is paired with a charge of opposite sign, or destabilization when it is in the

Table 5.1. Amino acid side chain pK_a values in solution and in enzyme active centres

Side chain	Free amino acid	Active centre*	Enzyme
Glu (γ-carboxyl)	3.9	6.5	Lysozyme
His (imidazole)	6.0	5.2, 6.8	Ribonuclease
Cys (thiol)	8.3	~4	Papain
Lys (ϵ-ammonio)	10.8	5.9	Acetoacetate decarboxylase

*These values are *not typical* but illustrate the magnitude of perturbations that can arise in special environments.

vicinity of a charge of the same sign. Thus, a carboxyl group juxtaposed with a positively charged lysine side chain should have a pK_a value lower than 'normal' because of stabilization of the carboxylate anion. The low pK_a values (~2.0) for the α-carboxyl groups of free amino acids result from the effect of the neighbouring protonated α-amino group.

Thus, even if we can determine the apparent pK_a of an ionizing group intimately associated with enzyme catalytic activity we must be cautious about assigning it to a particular kind of amino acid side chain. It is rarely, if ever, possible to be confident of such assignments unless there is independent evidence, such as chemical modification, X-ray structure or the effect of organic solvents on the pK_a values.

One must always ensure that pH-dependent denaturation is not a component of an observed pH-dependence of catalytic activity. The rate of denaturation is itself pH-dependent, so one must ensure that there is no significant denaturation during the time taken for an enzyme assay. This is easily tested by incubating the enzyme at each of several pH values spanning the range to be studied for the time required for an activity measurement, then returning the pH to a neutral value and assaying the enzyme. If all assays done in this way yield the same rate one can proceed to gather pH-dependent rate data over this pH range; if not, the pH range must be reduced to avoid any denaturation.

It is also important to note whether any change in the ionization of groups on the substrate is likely to occur over the pH range being studied. A change in the charge structure is likely to affect its ability to bind to the enzyme and may also affect the efficiency of catalysis, depending on the location of the charge in relation to the site of catalysis.

2. Sigmoid pH-dependence

Although it is widely known (and mentioned in most biochemical textbooks) that enzymes often display a bell-shaped dependence of catalytic activity on pH, it is instructive to begin by considering a simpler type of behaviour in which a sigmoid curve is observed. This is well illustrated by the proteolytic enzyme

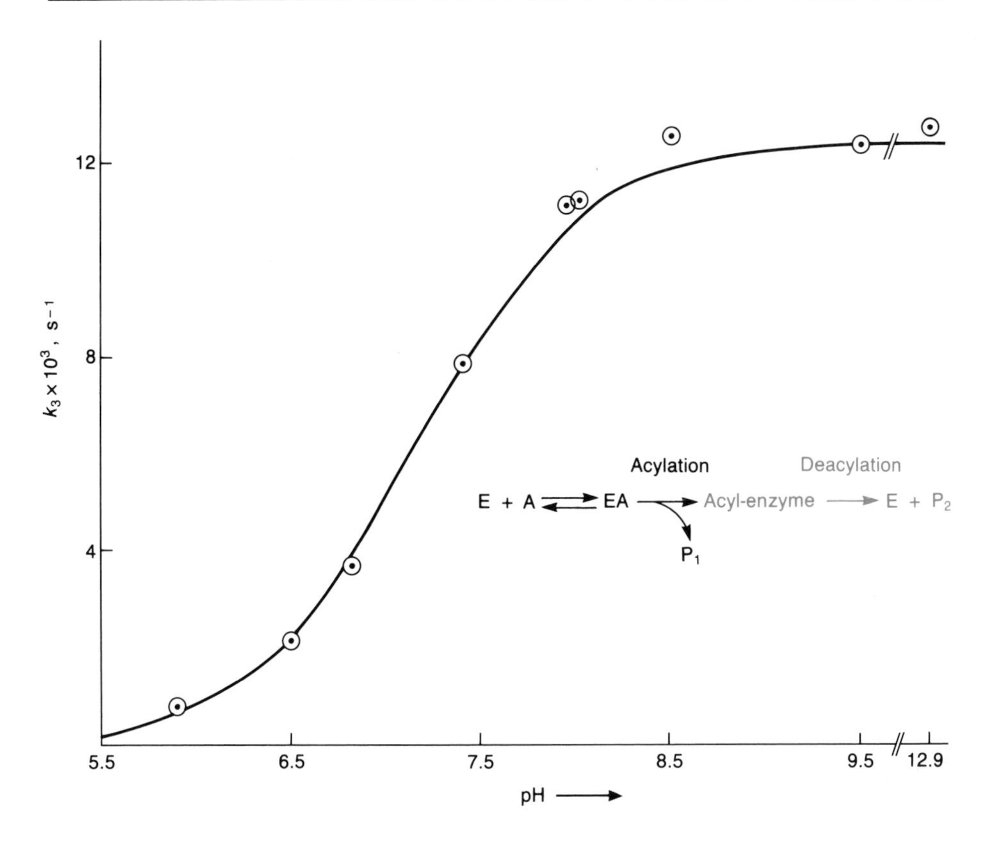

Figure 5.1. The pH-dependence of the deacylation of *trans*-cinnamoyl-chymotrypsin. This is the second part (colour) of the reaction scheme shown in the inset.

α-chymotrypsin (1), which catalyses the hydrolysis of peptide bonds usually adjacent, on the carboxyl-terminal side, to amino acids with aromatic side chains. The reaction proceeds in two stages, as shown in the inset to *Figure 5.1*. A serine residue, Ser-195, is acylated by the amino-terminal portion of the peptide or protein and the carboxyl-terminal portion is released into solution in the first ('acylation') step of the reaction. Subsequently the acyl-enzyme intermediate is hydrolysed to give the amino-terminal product ('deacylation'). Chymotrypsin can also be acylated by a number of synthetic substances that vary in their resemblance to a natural (protein) substrate. *Figure 5.1* shows the pH-dependence of the deacylation of *trans*-cinnamoyl-chymotrypsin (i), which forms in the course of the hydrolysis of *trans*-cinnamate esters and *trans*-cinnamoylimidazole (ii) (*Scheme 5.1*).

The enzyme requires one or more groups to be deprotonated for deacylation to occur, as is evident from the fact that the rate decreases to zero at low pH. In the simplest model to account for such behaviour the proton acts as an uncompetitive inhibitor of the reaction (Chapter 4, Section 3); a deprotonated complex can react but a protonated complex cannot. By analogy to Equation

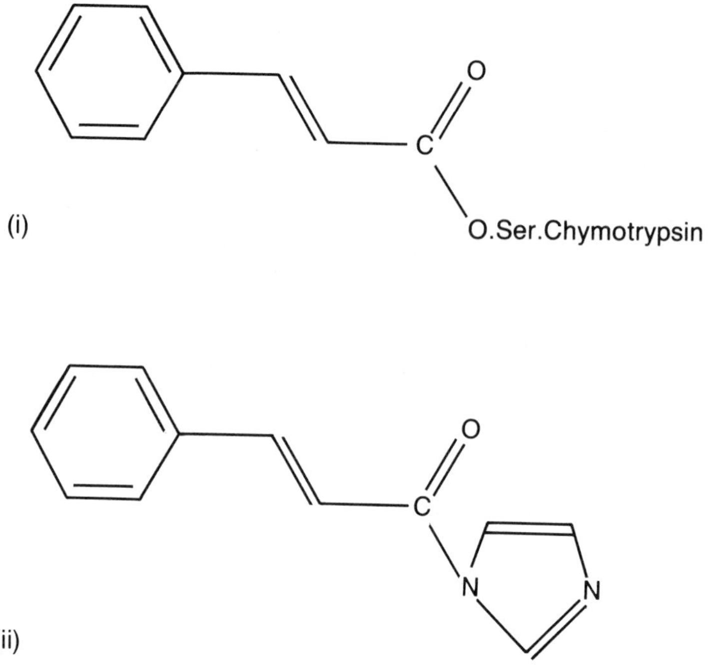

(i) O.Ser.Chymotrypsin

(ii)

Scheme 5.1.

4.8, therefore, the observed rate constant k_{obs} for deacylation is:

$$k_{obs} = k/(1 + h/K_a) \tag{5.1}$$

where h represents hydrogen ion concentration (a convention that will be followed throughout this chapter), k is the rate constant for the deprotonated acyl-enzyme, and K_a is the acid dissociation constant, which corresponds to the uncompetitive inhibition constant in Equation 4.7.

This equation can be rearranged to show that a plot of $1/k_{obs}$ against h is a straight line with a slope of $1/kK_a$ and intercept of $1/k$ on the ordinate:

$$1/k_{obs} = 1/k + h/kK_a \tag{5.2}$$

So the rate constant k for the deprotonated complex and the dissociation constant K_a can be determined from such a plot. For the data of *Figure 5.1* the value of the pK_a for deacylation of acyl-chymotrypsin proves to be close to 7.0. Considering this value in relation to *Table 5.1*, it is evidently appropriate to postulate that the observed ionization may be that of a histidine residue in the enzyme. Conceivably a carboxyl group could be responsible, but so great a perturbation would be unprecedented.

The X-ray structure of α-chymotrypsin reveals that the active site includes three amino acid side chains, those of Asp-102, His-57 and Ser-195, arranged as we shall discuss in Section 2 of Chapter 6. As Ser-195 is acylated to form the acyl-enzyme (His-57 being then bonded to water), both histidine and aspartate residues must exist close to the site where bond formation and cleavage occur in the deacylation reaction. Given that hydrogen-bonding can perturb pK_a values, how can one be certain that the pK_a of 7 is the result of His-57 and not Asp-102? It has been argued that the hydrogen bonds at the active centre have balancing effects on the pK_a of His-57, which in consequence displays a value of 7, close to the free-solution value for histidine. The bulk of independent evidence (e.g. NMR and chemical modification) supports the assignment of the pK_a to His-57, but there are one or two experiments (e.g. IR spectroscopy to monitor the ionization of Asp-102) that would favour assignment of the pK_a to Asp-102.

This illustrates the ambiguities of assignment that arise even in the relatively simple case of ionization of a single group in a well-characterized enzyme. Far greater difficulties may await those who try to interpret the more common bell-shaped pH-dependences.

3. Acid – base catalysis

A bell-shaped pH-dependence (*Figure 5.2*) implies that (at least) two ionizing groups change their state of protonation as the pH is raised from a value at the acid end where the enzyme is essentially inactive. Catalytic activity requires one of these groups to be deprotonated while the other retains its proton. The enzyme thus requires the simultaneous presence of the acid and the base to promote electron flow via the substrate(s) that results in covalent bond exchange and gives rise to products. The simplest realistic model for a bell-shaped pH-dependence is then as follows:

$$\begin{array}{ccccc} EH_2 & & EH_2A & & \\ \uparrow\downarrow K_{E1} & & \uparrow\downarrow K_{EA1} & & \\ EH + A & \underset{k_{-1}}{\overset{k_1}{\rightleftharpoons}} & EHA & \xrightarrow{k_2} & EH + P \\ \uparrow\downarrow K_{E2} & & \uparrow\downarrow K_{EA2} & & \\ E & & EA & & \end{array} \qquad (5.3)$$

The need for acid – base catalysis requires that only the singly protonated enzyme – substrate complex (EHA) can give products, but does not imply anything about which forms of the enzyme can bind the substrate. If the protonation/deprotonation steps can be treated as equilibria (as is usually valid since proton exchange reactions in buffered solutions are very fast for the ionizing groups that occur in proteins), there need not be explicit steps for binding A to EH_2 or E in the model.

The rate equation may be derived by the methods of Chapter 2 and is:

$$v = k_0e_0a/[K_m(1 + h/K_{E1} + K_{E2}/h) + a(1 + h/K_{EA1} + K_{EA2}/h)] \qquad (5.4)$$

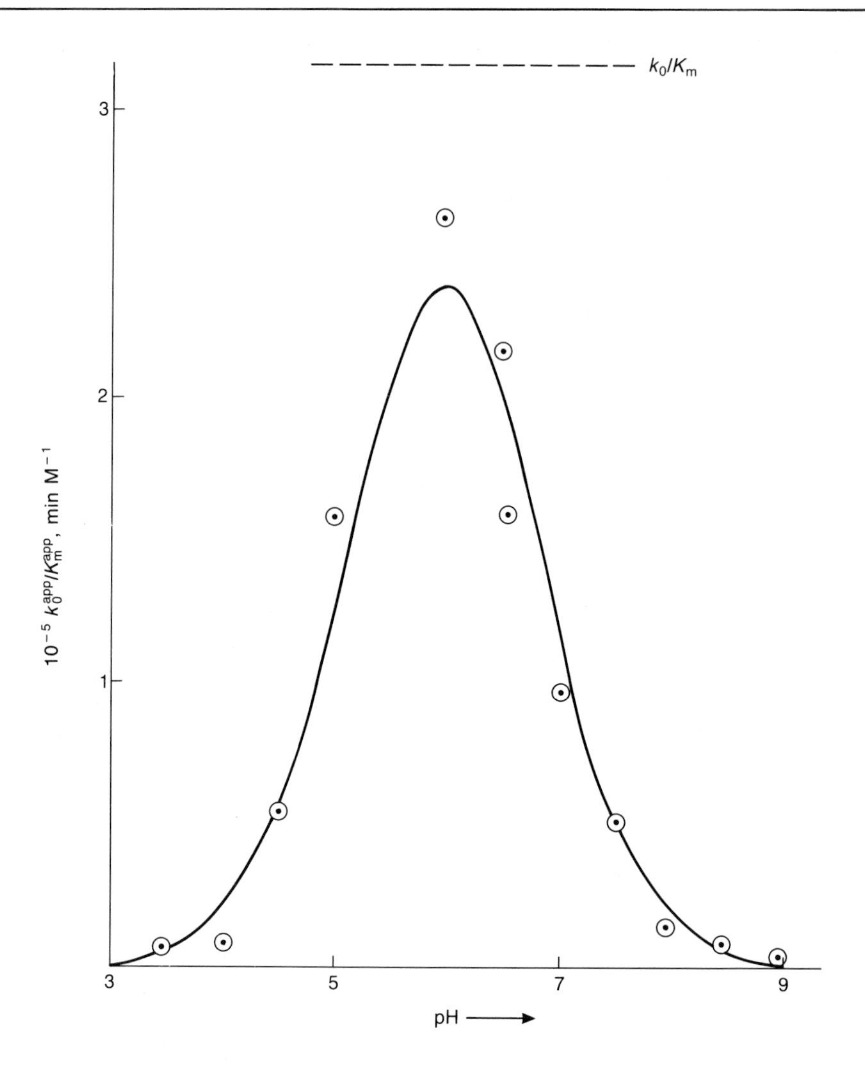

Figure 5.2. A bell-shaped pH-dependence. The data refer to the ribonuclease-catalysed hydrolysis of cytidine 2′,3′-cyclic phosphate (2).

where k_0, here equal to k_2, and K_m, which equals $(k_{-1} + k_2)/k_1$, are the so-called 'pH-independent' parameters of the reaction, that is the Michaelis–Menten parameters that would apply if the enzyme existed only in the singly protonated state. Dividing numerator and denominator by the term that appears in the denominator as the coefficient of a shows that Equation 5.4 has the form of the Michaelis–Menten equation:

$$v = k_0^{app} e_0 a / (K_m^{app} + a) \tag{5.5}$$

in which k_0^{app} and K_m^{app} are functions of the hydrogen ion concentration:

$$k_0^{app} = k_0/(1 + h/K_{EA1} + K_{EA2}/h) \quad (5.6)$$

$$K_m^{app} = K_m(1 + h/K_{E1} + K_{E2}/h)/(1 + h/K_{EA1} + K_{EA2}/h) \quad (5.7)$$

$$k_A^{app} = k_0^{app}/K_m^{app} = k_A/(1 + h/K_{E1} + K_{E2}/h) \quad (5.8)$$

These equations resemble Equations 4.11 – 13 for mixed inhibition. The appearance of terms in $1/h$ in addition to those in h just reflects the fact that we are assuming two sites for proton binding whereas in the inhibition case we assumed just one site for inhibitor binding.

To interpret Equations 5.6 – 8 properly, note that k_0^{app}/k_0 is the fraction of the enzyme – substrate complex in the singly protonated, catalytically competent state, and k_A^{app}/k_A is the fraction of free enzyme in the singly protonated state. The pH-dependence of the apparent K_m is more complex as it involves both factors. It follows that one can study ionization of the free enzyme by measuring the apparent value of k_A as a function of pH, whereas one can study ionization of the enzyme – substrate complex by measuring the apparent value of k_0 as a function of pH. This implies a considerable amount of work, but one should avoid the temptation to decrease it by measuring initial rates as a function of pH at only two substrate concentrations, one low, to mimic the behaviour of k_A, and one high, to mimic that of k_0. This is dangerous, because variation of apparent K_m with pH can lead to quite misleading results.

4. Determination of pK_a values

If the two pK_a values which define a bell-shaped curve are well separated (by at least 3 units, corresponding to a ratio of more than 1000 in the corresponding dissociation constants), the curve defined by Equation 5.6 or 5.8 as a function of pH shows a half-maximal value of the appropriate parameter when the pH is equal to one or other pK_a value. Each half of the curve may then be analysed as a sigmoid curve as in Section 2. However, *this is not the usual situation*, and normally the two pK_a values are too close for this sort of analysis to be valid.

Figure 5.3 shows a family of bell-shaped curves for various degrees of separation of pK_a values. The maximum value of the function progressively decreases as the pK_a values approach one another. When they are equal, the fraction of enzyme in the singly protonated state has a maximum of only one-third.

The most accurate methods of estimating pK_a values from bell-shaped curves involve curve-fitting computational techniques (3) outside the scope of this book. A less accurate but commonly used alternative uses plots of the logarithm of the measured parameter against pH, but this needs considerable care to give acceptable results, and becomes less accurate as the pK_a values become closer together. The most reliable simple method is to use the following relationship

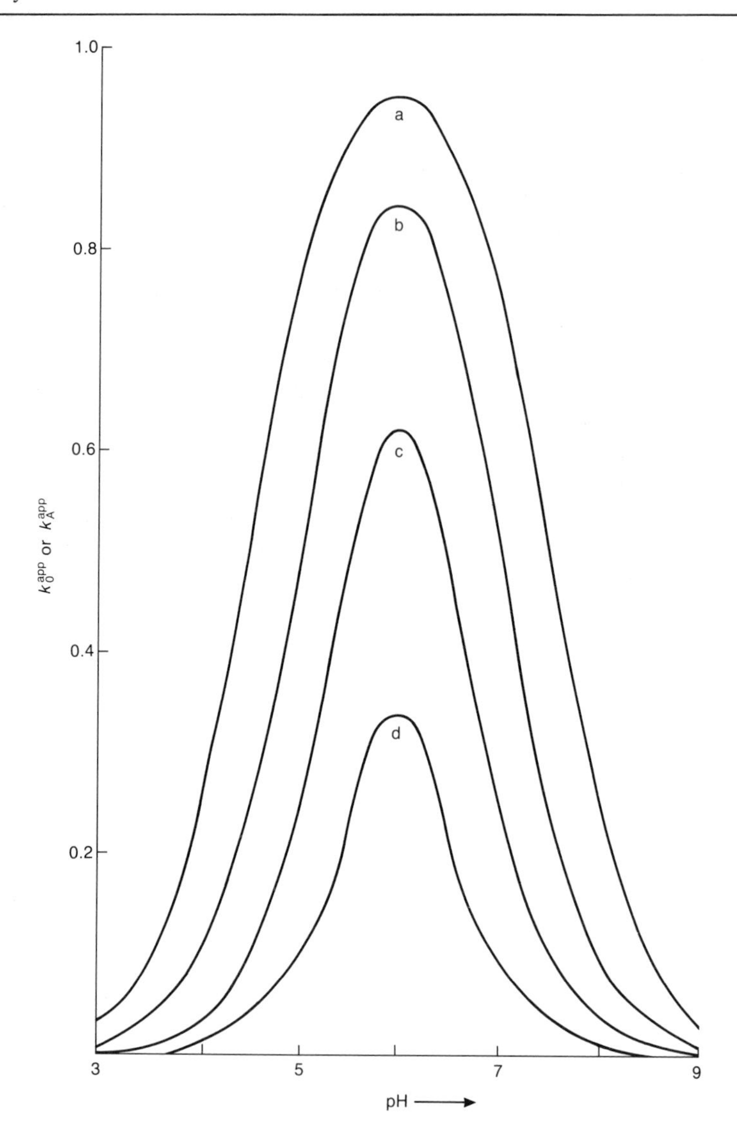

Figure 5.3. Theoretical bell-shaped curves. The curves are calculated from either Equation 5.6 or 5.8 with pK_a values as follows: (a) 4.5 and 7.5; (b) 5.0 and 7.0; (c) 5.5 and 6.5; (d) 6.0 and 6.0.

(4) between the difference between pK_a values and the width w of the curve at the half-height:

$$\text{p}K_2 - \text{p}K_1 = 2 \log (10^{w/2} - 4 + 10^{-w/2}) \tag{5.9}$$

As the pH at the maximum is the mean pK_a, that is (pK_1 + pK_2)/2, regardless of the difference between pK values, the individual values may easily be calculated from the pH values at the maximum and half-maximum values.

5. Interpretation of pK_a values

One might hope that, after coping with the complexities of environmental perturbation of pK_a values and estimating them from bell-shaped curves, their interpretation would be straightforward. There is still, however, an elusive factor to be taken into account. For a dibasic acid with two ionizable groups, α-H and β-H, there is no justification for assuming that α-H must always ionize first (even if it has the higher dissociation constant). In reality the singly protonated form of the dibasic acid will be a mixture of two species (*Figure 5.4*). The dissociation constants $K_{\alpha 1}$ and $K_{\alpha 2}$ both refer to ionization of α-H, but they are not in general equal because the ease of dissociation varies with the charge of the molecule from which it is dissociating; in the same way $K_{\beta 1}$ and $K_{\beta 2}$ both refer to dissociation of β-H, and are also in general unequal. These dissociation constants are called *microscopic* or *group* dissociation constants, and are the quantities of real interest. They are not, however, the quantities that can readily be measured, because titration methods do not normally distinguish between species of the same charge. Thus H-$\alpha\beta^-$ and $^-\alpha\beta$-H are counted together in defining the *molecular* dissociation constants K_1 and K_2:

$$K_1 = [(\text{H-}\alpha\beta^-) + (^-\alpha\beta\text{-H})]h/(\text{H-}\alpha\beta\text{-H}) \quad (5.10)$$

$$K_2 = (^-\alpha\beta^-)h/[(\text{H-}\alpha\beta^-) + (^-\alpha\beta\text{-H})] \quad (5.11)$$

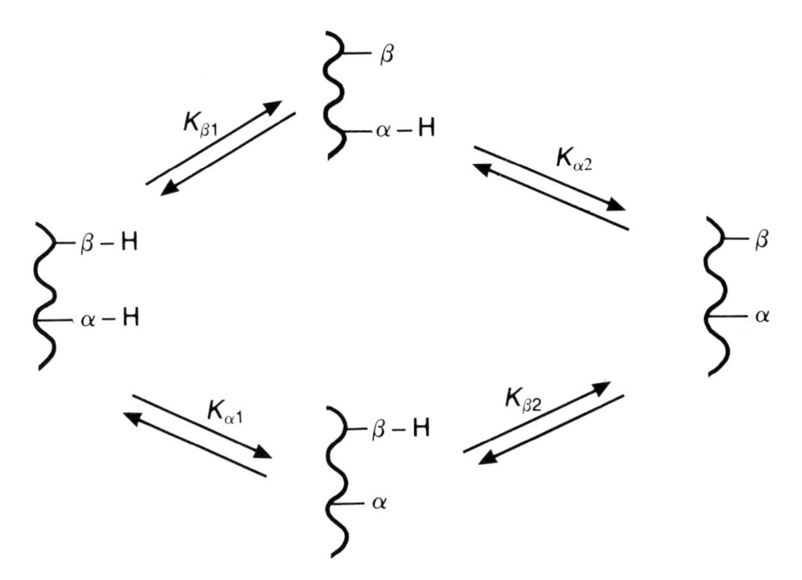

Figure 5.4. Scheme for the dissociation of an unsymmetrical dibasic acid. Note that the singly ionized form is a mixture of two species.

Comparing these with the definitions of the microscopic constants it is easy to show that they are related as follows:

$$K_1 = K_{\alpha 1} + K_{\beta 1} \qquad (5.12)$$

$$1/K_2 = 1/K_{\alpha 2} + 1/K_{\beta 2} \qquad (5.13)$$

K_1 refers to the first stage of ionization (without regard to route) and K_2 refers to the second stage. The ratio of concentrations of the two singly ionized forms is equal to the ratio of microscopic dissociation constants:

$$(^{-}\alpha\beta\text{-H})/(\text{H-}\alpha\beta^{-}) = K_{\alpha 1}/K_{\beta 1} = K_{\alpha 2}/K_{\beta 2} \qquad (5.14)$$

To characterize the pH-dependence of an enzyme fully we should like to know this ratio, but unfortunately it cannot be obtained from kinetic measurements; all one can obtain are estimates of the molecular dissociation constants K_1 and K_2. It follows that we cannot in practice determine which group ionizes first, and thus we do not know which group acts as an acid in the catalysis and which acts as a base. Even if we could be certain of the major component of the mixture of singly protonated species, we could not be sure it was responsible for the catalytic activity: the minor species might be much more reactive, to an extent that compensated for its lower concentration.

Additional information is required to resolve this uncertainty, and many physical techniques, such as potentiometric titration, UV and visible spectroscopy, and NMR, have been used in attempts to measure the pK_a values of catalytic groups. All of these methods suffer from drawbacks, some of them common to all methods. It is generally difficult to focus the titration method on the group of interest in the enzyme active centre, and efforts to do so frequently perturb the system.

Because of mutual effects between ionizing groups in close proximity to one another one must obtain at least one of the microscopic constants in conditions where the other cannot ionize (so that its changed charge during titration does not perturb the titrated group). This usually requires chemical modification, but this is, in itself, a perturbation of the system that introduces a new unknown.

This problem is of considerable theoretical importance, but no fully satisfactory solution to it has been found or seems likely to be found in the immediate future. The approach that has been adopted up to the present has been to apply as many different techniques as possible in the hope that consensus will emerge. This has happened in some cases but the results must still be regarded with scepticism.

6. Examples of pH-dependence studies

The example of a bell-shaped pH-dependence shown in *Figure 5.2* referred to the ribonuclease-catalysed hydrolysis of cytidine 2′,3′-cyclic phosphate (2). This enzyme catalyses the hydrolysis of RNA in a two-step process involving a cyclic phosphate intermediate. *Figure 5.2* shows the pH-dependence of the second step, the rate-limiting step of the reaction as a whole. The pK_a values of 5.2 and 6.8 are those of groups on the free enzyme and were assigned, before the X-ray structure was determined, to two histidine residues. Extensive chemical modification studies (alkylation of the enzyme by iodoacetate) and subsequent determination of the X-ray structure of the enzyme as well as that of an enzyme – inhibitor complex, support these assignments. In the formation of the cyclic intermediate His-12 acts as a base and His-119 as an acid, these roles being reversed in the hydrolysis of the cyclic intermediate. The lower pK_a of 5.2 for His-119 is nearly two units below that of a normal imidazole group in solution (see *Table 5.1*).

Another example is provided by the pH-dependence of the specificity constant for the α-chymotrypsin-catalysed hydrolysis of *N*-acetyltryptophan ethyl ester (5), which is bell-shaped with pK_a values of 6.8 and 8.8. The pK_a of 6.8 has been assigned to His-57, which is required to be in the base form for catalytic activity (see Chapter 6, Section 2 and Section 2 of this Chapter). This pK_a value is close to the value in free solution, surprisingly so in view of the complex hydrogen-bonding environment of His-57. The higher pK_a value has been assigned to the *N*-terminal Ile-16 α-amino group. When protonated, this group forms a buried salt bridge with the negative charge of Asp-194, an interaction that is essential for maintaining the substrate-binding site. Thus, the upper limb of the bell-shaped curve refers to an effect on K_m that is not directly related to the catalytic component of the mechanism. This example shows that one must be cautious about making a simple interpretation in terms of acid – base groups involved in catalysis: any ionizing group that affects the ability of the enzyme to bind substrate will appear in the pH-dependence of the specificity constant.

In conclusion, despite the many pitfalls and ambiguities, studies of pH-dependence have proved to be among the most incisive methods for studying enzyme mechanisms. As always, it is dangerous to interpret the results of experiments in isolation, but provided that all relevant chemical and enzymological evidence is taken into account, serious errors of interpretation are unlikely. Studies of pH-dependence have been very valuable for suggesting possible amino acids that may be implicated in the mechanism, which can be tested afterwards by chemical modification or in other ways.

7. Further reading

Boyde,T.R.C. (1980) *Foundation Stones of Biochemistry*. Voile et Aviron, Hong Kong. (Excellent both on the meaning of pH in relation to physical chemistry and on the origins of the concept.)

Tipton,K.F. and Dixon,H.B.F. (1979) *Methods Enzymol.*, **63**, 183.

Wharton,C.W. and Eisenthal,R. (1981) *Molecular Enzymology*. Blackie, Glasgow.

8. References

1. Bender,M.L., Schonbaum,G.R. and Zerner,B. (1962) *J. Am. Chem. Soc.,* **84**, 2562.
2. Herries,D.G., Mathias,A.P. and Rabin,B.R. (1962) *Biochem. J.,* **85**, 127.
3. Dixon,H.B.F. (1979) *Biochem. J.,* **177**, 249.
4. Wilkinson,G.N. (1961) *Biochem. J.,* **80**, 324.
5. Bender,M.L., Clement,G.E., Kezdy,F.J. and Heck,H.d'A. (1964) *J. Am. Chem. Soc.,* **86**, 3680.

6

Enzyme mechanisms

1. Introduction

In contrast to the rather theoretical emphasis of the earlier chapters, we shall now consider how enzyme kinetic analysis has contributed to our understanding of the chemical mechanisms of enzyme catalysis. In addition, we shall indicate what extra information, in terms of structure determination etc., is needed for deducing a chemical mechanism from an empirical kinetic study.

Before discussing chemical mechanisms of enzyme catalysis it is useful to summarize the ultimate objectives of enzyme kinetic analysis. Firstly, determination of an empirical reaction scheme defines the stoicheiometries and the nature of the kinetic mechanism, for example whether a reaction of two substrates requires them to bind in a particular order (Chapter 3). Secondly, provision of a quantitative description of enzyme catalysis allows the prediction of rates in general and identification of the rate-limiting step(s) in particular. This will normally require measurement of individual rate constants by fast reaction studies. Thirdly, one needs to check the kinetic competence of a putative chemical mechanism: is it consistent with the kinetic observations? Fourthly, one must seek direct evidence for a proposed mechanism, for example identification of a catalytically essential amino acid side chain by analysis of pH-dependence (Chapter 5). Lastly, assessment of the likely physiological (metabolic) role of an enzyme requires extra information in terms of the concentrations *in vivo* of the substrate, product, coenzyme and enzyme, as well as an analysis of inhibition and activation properties.

These objectives illustrate the fundamental nature of the information that can be derived from kinetic studies. A prerequisite for specific mechanistic studies is a thorough study of both the pre-steady-state and steady-state kinetics. The additional information that is needed for converting a kinetic mechanism into a chemical mechanism will be illustrated by reference to α-chymotrypsin, discussed already in Chapter 5, as it is probably the most thoroughly studied of all enzymes.

2. The mechanism of α-chymotrypsin

Many chemical reagents react more or less specifically with particular amino acid side chain groups in proteins. Inhibition of an enzyme by such a reagent suggests the involvement of the relevant amino acid in the catalytic activity, though structural effects that may result from modification of groups remote from the catalytic centre can be difficult to eliminate. Fortunately, groups at the active centre often show enhanced reactivity towards modifying reagents, and so it is not unusual for complete inactivation of an enzyme to result from reaction with a single amino acid. A good example of this is seen in the inhibition of the serine proteinases, including α-chymotrypsin (1), by diisopropyl fluorophosphate, as shown in *Scheme 6.1*.

Scheme 6.1. ^iPr = isopropyl.

Radioactive labelling of the reagent followed by analysis of the labelled protein allows the single amino acid that has reacted to be identified; in the case of α-chymotrypsin it is Ser-195.

The active site-directed inhibitor *p*-tosyl-L-phenylalanine chloromethylketone reacts with His-57 at the active centre of α-chymotrypsin and causes complete inactivation (*Scheme 6.2*) (2). This reagent has the advantage of possessing some

Scheme 6.2. Im = imidazole group.

characteristics of the substrate of α-chymotrypsin, which preferentially catalyses the hydrolysis of bonds adjacent to residues with aromatic side chains. Consequently it binds at the active centre, where an enhanced reaction occurs.

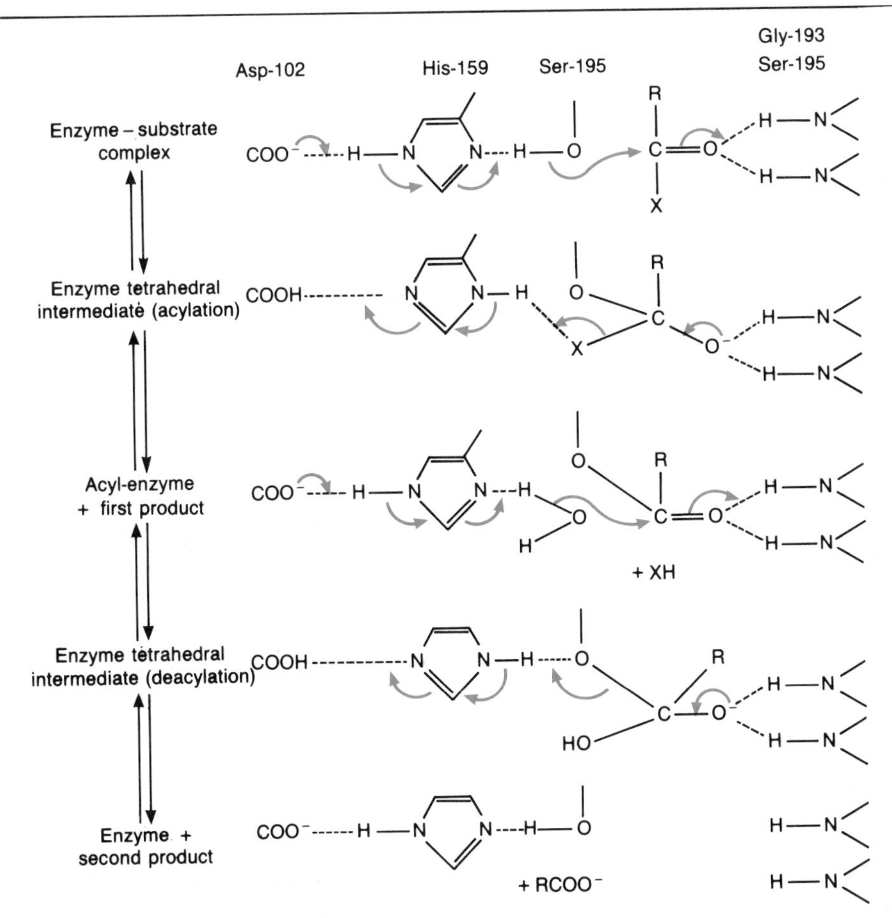

Figure 6.1. Postulated mechanism for α-chymotrypsin-catalysed hydrolysis. The substrate is represented as RCOX.

Use of these two reagents has thus provided evidence for the involvement of Ser-195 and His-57 in the enzyme mechanism; the pH-dependence of this enzyme (Chapter 5, Section 6) also supports a role for a histidine residue.

X-Ray crystallography has been crucial in the elucidation of enzyme mechanisms, as it has provided a framework in which the structures of reaction centres can be deduced, by determination of the structures of enzyme-bound inhibitors and products. By a combination of chemical reactivity data, pH-dependences and direct structural information, one can construct a picture of substrate binding and rationalize the abnormal chemical reactivity of active-centre groups in terms of their local environment. One of the biggest surprises in the early days of these studies was that catalytic systems often comprise several groups, hydrogen-bonded to one another and capable of acting in concert.

The best known such system occurs at the active centre of α-chymotrypsin (3). It consists of the side chains of Ser-195, His-57 and Asp-102 arranged as

Table 6.1. Catalytic constants (k_0) for the α-chymotrypsin-catalysed hydrolysis of ester and amide substrates

Substrate		k_0 (s^{-1})
Acyl portion	Leaving group	
N-Acetyl-L-tryptophan	–amide	0.026
	–L-alaninamide	2.8
	methyl ester	28
	p-nitrophenyl ester	30

in *Figure 6.1*. The existence of an interacting system of such complexity could not have been inferred from reactivity data alone. Moreover, its discovery created an entirely new problem, as classical chemistry could offer little insight into the prediction and interpretation of assemblies of groups of this type. In an important sense, therefore, the study of enzyme mechanisms acquired its own identity as a branch of catalysis, as it was apparent that a simple extension of solution chemistry would not provide answers to the problems of how enzymes work.

Inspection of the kinetic data for α-chymotrypsin (4) collected in *Table 6.1* reveals several important points relevant to the mechanism. Firstly, simple amide substrates are hydrolysed more slowly than comparable esters. Secondly, the large increase in chemical reactivity accompanying a change in the leaving group of ester substrates from methyl to *p*-nitrophenyl does not significantly affect the catalytic constant, which can only be explained easily by supposing that the two substrates are hydrolysed with the same rate-limiting step. The additional information that a 'burst' of *p*-nitrophenol, stoicheiometric with the enzyme concentration, is released on mixing the enzyme with a *p*-nitrophenyl ester substrate (5), but not with an amide substrate, shows that the mechanism must be defined as a two-step process (*Scheme 6.3*).

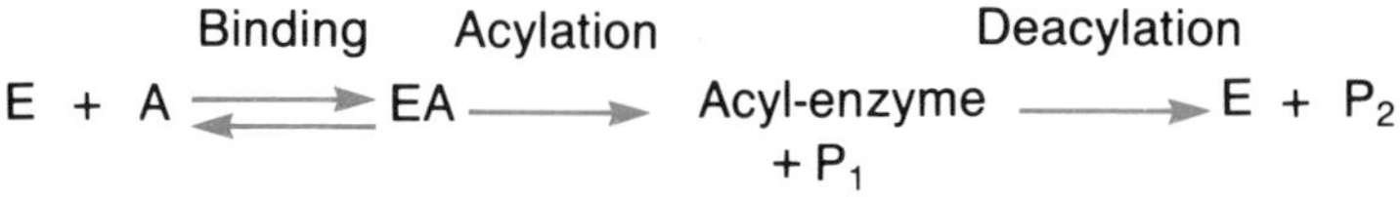

$$E + A \underset{}{\overset{\text{Binding}}{\rightleftharpoons}} EA \xrightarrow{\text{Acylation}} \text{Acyl-enzyme} + P_1 \xrightarrow{\text{Deacylation}} E + P_2$$

Scheme 6.3.

The data in *Table 6.1* are now interpreted by proposing that for amides k_2 is small compared with k_3, whereas the reverse is true for esters. Using ^{14}C-labelling of the acyl portion of the substrate and isolation of the intermediate at low pH, it has been shown that the intermediate acyl-enzyme is an ester of Ser-195. That the K_m values of the two amide substrates are very similar indicates the involvement of strain in the mechanism; all of the potential binding energy of the alanine residue is converted into catalytic rate enhancement.

An important rate-enhancing feature of the mechanism identified by kinetic and structural studies is the 'oxyanion hole' interaction. The carbonyl oxygen of the susceptible peptide or ester bond forms two hydrogen bonds with protons attached to main-chain amide nitrogens (6). This stabilizes the transition states of the reaction by encouraging accumulation of negative charge on the oxygen atom.

The information presented here may be combined with much other evidence to propose the mechanism for α-chymotrypsin shown in *Figure 6.1*. Many textbooks show Asp-102 as playing only a passive role in the mechanism with no change in its protonic state during a catalytic cycle. Recent experiments using ^2H_2O as solvent instead of ordinary water have shown that *two* protons move when the transition states form (7). This requires Asp-102 to act as an acid – base, not simply to provide electrostatic stabilization for the positively charged His-57.

The nature of the covalent-bond-exchange processes in the mechanism is now clearly defined, but little is known about the influence of the weaker non-covalent forces, such as hydrogen bonds and hydrophobic and electrostatic interactions. For example, the quantitative effect of the oxyanion hole interaction is unknown and difficult to measure by conventional techniques. These vital aspects of catalysis in which modulation of covalent-bond-exchange occurs are currently being studied by means of novel computational and spectroscopic methods, such as NMR. We shall not be able to claim that we 'understand' enzyme catalysis until we can define the influence of these forces in the extremely complex environment of the enzyme active site (with substrate bound). Chemistry has supplied ground rules for interpreting simple systems, but these need to be diversified and extended before enzyme catalysis can be fully explained.

3. Proof-reading in protein synthesis

The efficiency and survival of an organism require that it can synthesize proteins with extremely few errors in their amino acid sequences, because protein molecules with sequence errors will generally be less active as a result of incorrect folding, even if they contain no errors in the catalytic centre itself. These proteins are likely to be degraded rapidly and to impose a high cost to the energy supply of the cell.

tRNA molecules are charged with the appropriate amino acids by a series of enzymes known as aminoacyl-tRNA synthetases. The general kinetic mechanism of these enzymes involves the intermediacy of enzyme-bound aminoacyl-adenylates, which have to react with the appropriate kinds of tRNA:

$$\text{E} + \text{Val} + \text{ATP} \longrightarrow \text{E.Val–AMP} + \text{PP}_i$$

$$\text{E.Val–AMP} + \text{tRNA}^{\text{Val}} \longrightarrow \text{E} + \text{Val–tRNA}^{\text{Val}} + \text{AMP}$$

In this example tRNA^{Val} represents the particular kind of tRNA that recognizes codons for valine.

To avoid errors in protein synthesis each kind of tRNA must be charged with the right amino acid. For example, threonine, which has a very similar size and shape to valine, must not be charged onto $tRNA^{Val}$ to give Thr – $tRNA^{Val}$ molecules, or, if they are formed, these errors must be recognized and the wrongly charged tRNA destroyed before they are used in protein synthesis. An ingenious 'double sieve' editing system has evolved (8) that ensures a low error rate, though it does have some cost in energy. Two different effects are exploited, namely steric exclusion and ineffective binding. Steric exclusion works by preventing the binding of an amino acid side chain more bulky than the correct one; side chains smaller than normal can bind, but do so more weakly than the correct one because they lack groups used in generating binding energy. Charging of tRNA molecules by amino acids that are too big is virtually eliminated, whereas charging with ones that are too small may still occur with a frequency of 1 – 2%.

Such an error rate is, however, too great to be tolerated: for example, if each residue in a 100-residue protein has only a 0.98 chance of being correct, then the proportion of correct molecules produced is 0.98^{100}, that is 13% of the total. A second check is thus required and this is provided by a second active site on the enzyme distinct from that which catalyses the synthetic reaction; this has a hydrolytic activity that is used to hydrolyse incorrectly formed aminoacyl-adenylate and aminoacyl – tRNA. This site is smaller than the synthetic site and so excludes the correct amino acid side chain, but mischarged tRNA with smaller side chains can bind and are hydrolysed. In this way, the editing mechanism improves the fidelity of protein synthesis by a factor of around 10^3 at a relatively small cost in ATP. The value of this improvement may be judged by noting that 0.99998^{100} is 99.8%.

Amino acids that are *isosteric* (having the same size and shape), for example valine and threonine, present more of a problem as size discrimination cannot be used for the basis of a sieving mechanism. Thus, the enzyme specific for valine must exploit some other property to avoid mischarging with threonine. This can be achieved by providing a wholly hydrophobic binding site for the synthetic reaction but one with a hydrophilic region for the hydrolytic reaction. In this way valine binds tightly at the synthetic site but very weakly at the hydrolytic site, whereas the reverse is true for threonine. *Figure 6.2* illustrates the editing system for the enzyme specific for isoleucine. Not all of the aminoacyl-tRNA synthetases require a double-site editing mechanism; for example, the enzyme specific for tyrosine can achieve adequate discrimination without it.

4. Site-directed mutagenesis

The manipulation of substrate structure has allowed enzymologists to probe the nature of enzyme specificity and reactivity by kinetic and spectroscopic analysis. The mechanistic role of particular amino acids in an enzyme active centre may be studied directly by using *site-directed mutagenesis*, in which a given amino acid can be changed to one of the 19 others. To pinpoint enzyme – substrate

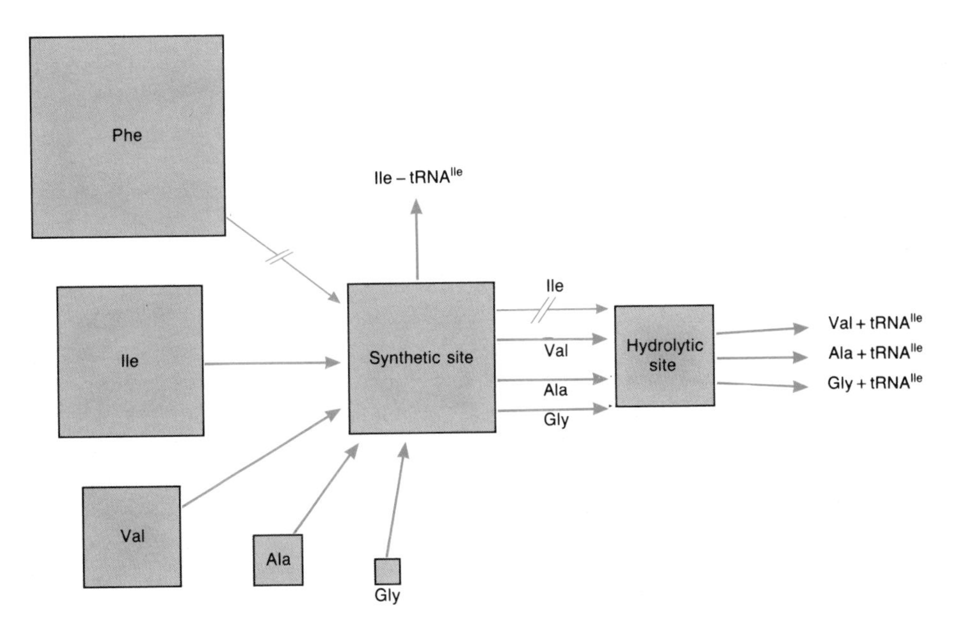

Figure 6.2. Double-sieve editing mechanism in aminoacyl-tRNA synthetases. In this illustration the 'correct' reaction is the formation of Ile – tRNA^{Ile}. However, smaller amino acids (Val, Ala, Gly) have some probability of forming an aminoacyladenylate, which is then hydrolysed at the hydrolytic site. The 'correct' intermediate Ile – AMP is too large for the hydrolytic site and is not hydrolysed. Amino acids larger than isoleucine (e.g. Phe) cannot undergo the first reaction. Possible reactions are shown in colour.

contacts it is clearly necessary to have a detailed knowledge of the enzyme active centre, which must be obtained from protein (or DNA) sequencing and X-ray crystallography. Determination of the structure of enzyme – inhibitor complexes also allows the investigator to make appropriate choices about which changes to make in the active centre of the enzyme.

Some of the most effective work of this type has been applied to the aminoacyl-tRNA synthetases (9) that were discussed in the previous section. First, the tyrosine-specific enzyme from a thermophilic bacterium was cloned and over-expressed in *Escherichia coli*, and mutants were made with modified K_m or k_0 values, reflecting changes in ability to bind substrate and to catalyse the reaction (see Chapter 1, Section 3), by replacing amino acids believed to participate in enzyme – substrate interaction. The pattern of contacts is shown in *Figure 6.3*, and *Table 6.2* lists kinetic constants for some of the mutants.

This type of work has shown that one can alter and sometimes increase the specificity of enzymes by site-directed mutagenesis. Thus, as well as being a powerful tool for analysing enzyme mechanisms and the source of enzyme catalytic power, it seems that it will be possible to 'engineer' new proteins for commercial or medical purposes. There is a rapidly growing interest in this field, reflecting the hope that new catalysts may be designed from pre-existing enzymes

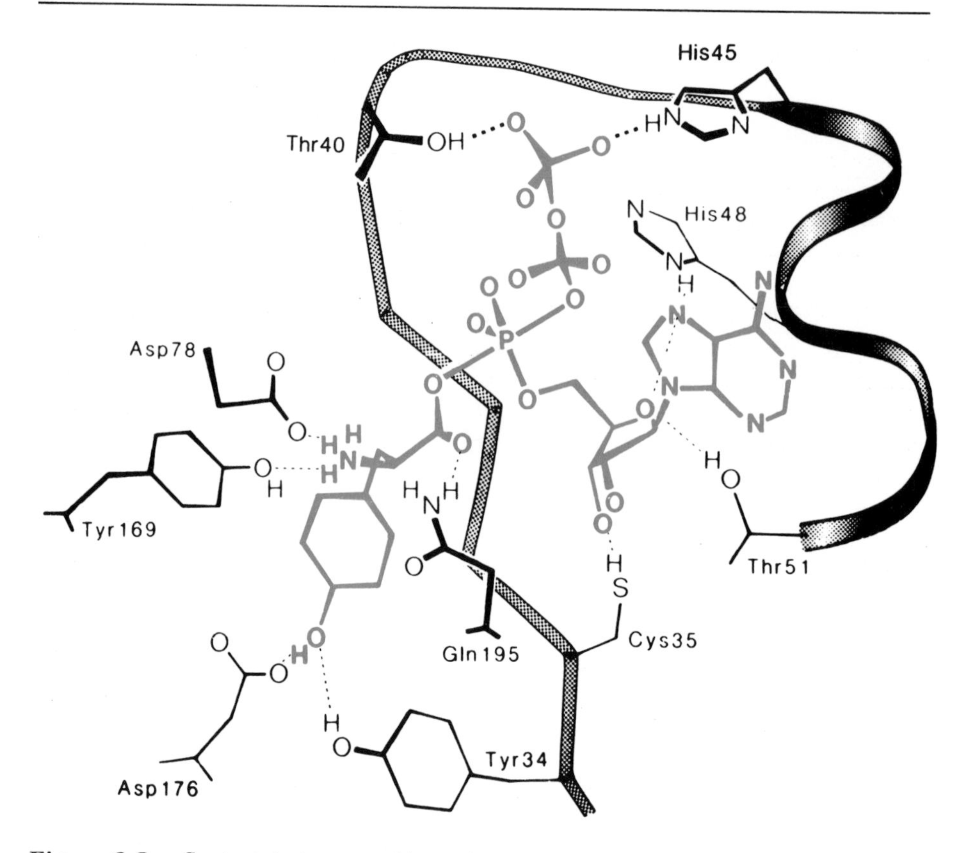

Figure 6.3. Contacts between residues of tyrosyl–tRNA synthetase and the transition state of ATP and tyrosine during the formation of tyrosyladenylate (9).

to tackle new tasks in biotechnology. For example, disulphide bonds have been introduced into proteins that do not naturally have them in the hope of increasing their thermal stability, but so far the improvements in stability obtained in this way have been marginal.

Table 6.2. Kinetic parameters for tyrosyl-tRNA synthetases derived from site-specific mutagenesis

Enzyme	k_0 (s^{-1})	K_m^{ATP} (mM)	k_{ATP} ($M^{-1}s^{-1}$)
Wild-type	4.7	2.5	1860
Thr-51→Ala-51	4.0	1.2	3200
Thr-51→Pro-51	1.8	0.019	95 800

These data of Wilkinson *et al.* (13) refer to the aminoacylation reactions catalysed by the wild-type enzyme and mutant enzymes with alanine or proline instead of threonine at position 51. Recall (from Chapter 1, Section 3) that the specificity constant k_{ATP} is the catalytic constant k_0 divided by the Michaelis constant K_m^{ATP}.

5. Triose phosphate isomerase: a highly efficient enzyme

Triose phosphate isomerase catalyses the interconversion of glyceraldehyde 3-phosphate and glycerone phosphate (often called 'dihydroxacetone phosphate'), which occurs in glycolysis. The reaction proceeds through an enediol intermediate with carboxyl (Glu-165) and imidazole (His-95) groups providing acid – base catalysis. It has a special claim on our attention because of the extensive studies of the energetics of the reaction and kinetic comparison with the corresponding reaction catalysed by acetate. The enzyme-catalysed reaction is many orders of magnitude faster than the acetate-catalysed alternative, and understanding how this is possible is valuable for understanding the nature of enzyme catalysis. We do not have space to discuss the mechanism in detail here, but will merely note some salient features (10).

Although the two reactions proceed through chemically similar intermediates, those in the enzyme-catalysed reaction are in general more stable than those in the acetate-catalysed reaction, and are formed much more readily. However, stabilization of intermediates must not be allowed to proceed too far, because they must not only be formed easily but must also continue easily to the next step. More important is that the energy barriers to their reaction, the transition states, should be as low as possible. The extent to which this is achieved by triose phosphate isomerase may be judged from the fact that the reaction rate is close to the *diffusion limit*: this expresses the fact that a reaction cannot proceed faster than the rate at which the reacting molecules find one another by diffusion through the solvent. Triose phosphate isomerase has been claimed as a 'perfectly evolved enzyme' because there is virtually no scope for further increases in the reaction rate as a result of further improvement in the structure. Whether one accepts this claim must depend, in part, on whether high reaction rates are the sole criterion of perfection: for many enzymes high specificity may be more important for life; for others a high degree of responsiveness to metabolic signals. For example, similar claims have been made for glutamine synthetase on the quite different grounds of its ability to recognize and respond to numerous physiologically important effectors (14).

6. Lactate dehydrogenase: protein mobility in enzyme catalysis

Lactate dehydrogenase catalyses the oxidation of lactate to pyruvate by NAD^+. X-Ray crystallography, together with kinetic and ligand-binding experiments, has suggested that it functions by an 'oil – water – histidine' mechanism (11). This involves a mobile loop of peptide that can close down over the enzyme site after both substrates are bound. NAD^+ binds first to the 'open loop' hydrophilic form of the active site; then lactate binds to the open site and induces closure of the loop. The environment changes from a water-based hydrophilic one to an 'oily' hydrophobic one in which the positive charge on NAD^+ loses its protective shield of solvent molecules, so that now a charge is in an energetically unfavourable hydrophobic environment. A histidine residue (His-195) acts as a base to remove the proton from the lactate hydroxyl group, promoting electron flow to neutralize the positive charge on NAD^+ and converting it to NADH.

The equilibrium constant for the complete reaction strongly favours lactate and NAD^+. However, the loop mechanism has the effect of reversing this for the enzyme-bound species in a hydrophobic environment, as a result of the loss of water around the charge, which increases the reaction rate. In consequence the dissociation of NADH from the enzyme becomes the rate-limiting step of the reaction.

Spectroscopy and kinetics have been used to very good effect in resolving events along the reaction pathway. Nitration of a tyrosine residue (Tyr-237) in the loop has allowed observation of the spectrum of this group in the near-UV region. As this spectrum is sensitive to the environment of the group it can be used to monitor whether the loop is open or closed. By making measurements under 'single-turnover conditions', where the enzyme concentration is made equal to the substrate concentrations, so that it behaves as a reagent rather than a catalyst, it has been possible to show that in the reverse direction the reaction conversion of the enzyme–NADH–pyruvate complex to give NAD^+ and lactate is limited by the rate of loop movement and not by hydride ion (negatively charged hydrogen atom) transfer.

Labelling of a cysteine residue (Cys-165) that is located between the active-site histidine residue and the loop, but which is apparently not directly involved in the reaction, with a specific ^{13}C-probe has allowed some ingenious experiments with NMR to study the rates of interconversion of the ternary enzyme–NADH–oxamate complex and the binary enzyme–NADH complex (12); oxamate is an analogue of pyruvate. Although ^{13}C-NMR measurements are rather slow, requiring some hours to accumulate an adequately noise-free signal, the *shapes* of the absorption lines depend on the lifetimes of the absorbing species. They can be used to obtain information about rates of interconversion if an equilibrium mixture is prepared in which there are significant amounts of both reactant and product forms. These experiments have shown that there is a conformational change with the same rate as independent kinetic measurements have shown to be rate-limiting during lactate synthesis from pyruvate. Such experiments illustrate how kinetic, structural and spectroscopic data can be united to suggest detailed mechanistic proposals that include aspects of protein mobility. Knowledge of the X-ray crystal structure of lactate dehydrogenase has permitted a much more detailed interpretation than we have given here.

7. Conclusions

In this chapter we have indicated some of the ways in which kinetic data have been used to define important mechanistic concepts. In the space available it is inevitable that there will be a gap between the fundamental kinetic properties described earlier in the book and the far-reaching deductions that have been based on a combination of kinetic, structural and spectroscopic information. Nonetheless, we hope that this brief exposure to more advanced ideas will encourage the reader to pursue the subject in more detailed sources.

It is always useful to have as clear a picture as possible of areas of a subject where new information is most urgently required, that is to understand what it is that we do not yet understand. Two aspects of enzyme mechanisms stand out at present. First, we know little of the nature, combination and distribution of the forces that allow enzyme–ligand interaction and transition-state stabilization. It has not yet been possible to define an evolutionary strategy in terms of enzyme–substrate contacts leading to optimal catalysis. There are some important guiding principles, such as transition-state stabilization, and some good experimental examples, such as triose phosphate isomerase (Section 5), but we cannot yet apply these principles in the general case to define an optimal protein structure.

Second is the role of protein mobility in enzyme catalysis, which has been elucidated in a few cases, such as lactate dehydrogenase (Section 6). It is likely that movement of amino acid side chains, and probably protein segments in many cases, is necessary in the catalytic cycles of enzymes. This has been deduced for some enzymes (e.g. rotation of His-159 in α-chymotrypsin) from studying the geometrical constraints of the chemical processes involved. In other cases, observations of rate-limiting conformational changes provide evidence for the importance of structural change in enzymes during catalysis.

Site-specific mutagenesis should allow further resolution of the roles of particular amino acids in secondary and tertiary structures of proteins. An extension of this will allow elucidation of the evolutionary strategy for choice of amino acids in active centres where there are contacts with substrates, as has already been done to some degree for the aminoacyl-tRNA synthetases (Section 4). Ideally one will be able to probe the geometric constraints of these choices as it seems that this may be the key to understanding catalytic rate enhancement.

There can be no doubt that kinetic studies will be central in future investigations of enzyme mechanisms, not least in the assessment of mutant enzymes. In view of the generally accepted importance of transition-state stabilization, it is as well to remember that the structures of real transition states (not analogues) can only be studied kinetically. It is equally clear, however, that only by combining the use of several techniques will the really important fundamental questions be answered.

8. Further reading

Fersht,A. (1985) *Enzyme Structure and Mechanisms*. Freeman, New York, 2nd edn.
Walsh,C. (1979) *Enzymatic Reaction Mechanisms*. Freeman, New York.
Wharton,C.W. and Eisenthal,R. (1981) *Molecular Enzymology*. Blackie, Glasgow.

9. References

1. Jansen,E.F., Nutting,M.D. and Balls,A.K. (1949) *J. Biol. Chem.,* **179**, 201.
2. Shaw,E. (1970) In Boyer,P.D. (ed.), *The Enzymes.* Academic Press, New York, 3rd edn., vol. 1, p. 91.
3. Blow,D.M., Birktoft,J.J. and Hartley,B.S. (1969) *Nature,* **221**, 337.
4. Zerner,B., Bond,R.P.M. and Bender,M.L. (1964) *J. Am. Chem. Soc.,* **86**, 3674.
5. Hartley,B.S. and Kilby,B.A. (1954) *Biochem. J.,* **56**, 288.
6. Steitz,T.A, Henderson,R. and Blow,D.M. (1969) *J. Mol. Biol.,* **46**, 337.
7. Stein,R.L. and Strimpler,A.M. (1987) *J. Am. Chem. Soc.,* **109**, 4387.
8. Fersht,A.R. (1977) *Biochemistry,* **16,** 1025.
9. Leatherbarrow,R.J., Winter,G. and Fersht,A.R. (1985) *Proc. Natl. Acad. Sci. USA,* **82**, 7840.
10. Albery,W.J. and Knowles,J.R. (1976) *Biochemistry,* **15**, 5631.
11. Clarke,A.R., Waldman,A.D.B., Hart,K.W. and Holbrook,J.J. (1985) *Biochim. Biophys. Acta,* **829**, 397.
12. Waldman,A.D.B., Birdsall,B., Roberts,G.C.K. and Holbrook,J.J. (1986) *Biochim. Biophys. Acta,* **870**, 102.
13. Wilkinson,A.J., Fersht,A.R., Blow,D.M. and Winter,G. (1984) *Nature,* **307**, 187.
14. Ottaway,J. (1988) *Regulation of Enzyme Activity—In Focus.* IRL Press, Oxford, in press.

Glossary

Although the definitions given here are as precise as we can make them consistent with conciseness, they should not be taken as a substitute for reading the text.

Apparent value: quantity that is constant only while specified concentrations are held constant.

Binary complex: complex formed by association of two species.

Catalytic constant: limiting rate divided by total enzyme concentration.

Cationic acid: cation (e.g. ammonium, NH_4^+) capable of donating a proton.

Competitive inhibition: type of inhibition characterized by decrease of the apparent value of the specificity constant but not that of the limiting rate.

Dead-end reaction: part of a mechanism that does not lead to product, which the enzyme can leave only by reversing the steps that brought it there.

Dead time: time between the start of a reaction and the first moment at which it can be observed.

Distribution equations: set of equations that define the fractions of enzyme existing as free enzyme and as each intermediate.

Elementary step: reaction consisting of a single step, or a single step in a mechanism.

Group-transfer reaction: reaction in which a chemical group is transferred from a donor to an acceptor.

Half-time: time for a reaction to go to half completion.

Haldane relationship: equation that relates the kinetic constants of an enzyme-catalysed reaction to its equilibrium constant.

Hydrogen bond: weak attraction between an electronegative atom (e.g. N, O) and the slight positive charge associated with a hydrogen covalently bound to another electronegative atom.

Hydrophobic interaction: apparent attraction, when dispersed in water, between molecules or groups that do not interact strongly with water.

Inhibition: tendency of an enzyme-catalysed reaction to proceed more slowly in the presence of a species called an inhibitor.

Isotope exchange: reaction catalysed by an enzyme when a label (usually radioactive) is transferred from one molecule to another involving no other chemical change.

Limiting rate: rate approached by an enzyme-catalysed reaction when the substrate concentrations become very large.

Maximum velocity: obsolete (and misleading) term for limiting rate (V).

Median: the middle of a set of values arranged in order.

Michaelis constant: substrate concentration at which the rate of an enzyme-catalysed reaction is half of the limiting rate.

Mixed inhibition: type of inhibition in which competitive and uncompetitive effects are both present.

Mutase: enzyme that catalyses an isomerization in which a phosphate group is moved from one position to another.

Order of reaction: power to which a concentration is raised in a rate equation.

Perturbation method: method of studying a rapid reaction by observing its behaviour after a sudden change in conditions.

Ping pong mechanism: alternative name for substituted-enzyme mechanism.

Primary plot: plot of kinetic data before any preliminary processing or interpretation.

Pseudo-first-order rate constant: product of a second-order rate constant and a constant concentration.

Quenched flow: method of studying a fast reaction by suddenly stopping it (e.g. with cold or by chemically destroying the enzyme) a short time after mixing.

Rate constant: quantity that when multiplied by the appropriate concentration or concentrations gives the rate of an elementary step of a reaction (or, loosely, the rate of a composite reaction treated as if it were an elementary step).

Saturation: tendency of the rate of an enzyme-catalysed reaction to approach a limit as the substrate concentration is increased.

Secondary plot: plot of the values obtained in preliminary analysis of a primary plot.

Serine proteinases: group of extensively studied enzymes (including α-chymotrypsin, trypsin, etc.) that catalyse the hydrolysis of proteins and protein fragments by a mechanism involving a reactive serine residue.

Specificity: capacity of an enzyme to discriminate between one substrate and another.

Specificity constant: kinetic constant in an enzyme-catalysed reaction obtained by dividing the catalytic constant by the Michaelis constant.

Steady state: state in which the concentration of each intermediate in a reaction remains constant because it is removed as fast as it is produced.

Stopped flow: method of studying a fast reaction by spectroscopic observation after rapid mixing of the reactants.

Substituted-enzyme mechanism: mechanism for a group-transfer reaction in which the group is first transferred to the enzyme.

Temperature jump: type of perturbation method in which an electric discharge is used to bring about a sudden increase in the temperature of a reacting mixture.

Ternary complex: complex formed by association of three species.

Uncompetitive inhibition: type of inhibition characterized by decrease of the apparent value of the limiting rate but not that of the specificity constant.

Index